The Enchantments of Technology

LEE WORTH BAILEY

University of Illinois Press

Urbana and Chicago

∞ This book is printed on acid-free paper.

Library of Congress Cataloging-in-Publication Data
Bailey, Lee Worth.
The enchantments of technology / Lee W. Bailey.
p.   cm.
Includes bibliographical references and index.
ISBN0-252-02985-2 (cloth : alk. paper)
ISBN0-252-07232-4 (pbk. : alk. paper)
1. Technology—History. I. Title.
T15.B2657   2005
600—dc22   2004029690

The Enchantments of Technology

CONTENTS

PREFACE

The powerful, positive and negative impacts of technology are the most pressing problems of our time. There are many interpretations of this process, and the power of technology to change and even destroy us compels us to examine these interpretations carefully. This book is about the enchantments of technology, or its imaginative, mythic, and unconscious dimensions, from the perspective of the humanities. My refections on this subject date back to my undergraduate days as an industrial design student at the University of Illinois, when I enjoyed learning to design technologies to fit human needs.

As my mind expanded in a more philosophical and spiritual direction, I was compelled to ask questions about the broader contexts of technology. Various new cultural crises, such as the assassination of John F. Kennedy and the reaction to Martin Luther King Jr.'s protests against racism, shoved aside my attraction to industrial design. I was compelled to study the ethical and spiritual depths of our culture behind these crises, so I went to graduate school to study Christianity, depth psychology, mythology, philosophy, and world religions. This book looks at technology through the eyes of an industrial designer and in light of these graduate studies, as well as my subsequent research and reflections.

I am grateful to my industrial design professors at the University of Illinois, who above all taught me to think creatively. I thank Don Ihde at SUNY–Stony Brook, whose outstanding work in the philosophy of technology inspired me to write this book, for his encouragement. I also thank Frederick Ferré, at the University of Georgia, for his excellent work in the philosophy of religion and the philosophy of technology and for his challenges and support. I am also grateful to Tom F. Driver at Union Theological Seminary at Columbia University, who awakened in me a sense of social justice and theological vision. I deeply appreciate David L. Miller at Syracuse University, who inducted me into the radically transforming insights of depth psychology, comparative mythology, and continental ontology. My methodology draws on but differs from each of theirs in various respects. I also gratefully thank Peter Fortunado in Ithaca, an accomplished poet and novelist, for valuable writing advice. I thank my wife and children abundantly for their patience with my time-consuming research and writing and their faith in the process. Special thanks goes to my editor at the University of Illinois Press, Kerry Callahan, who appreciated the "edge" in the manuscript and contributed valuable editorial advice. Thanks also to my friends for discussions, journal editors for publication of earlier versions of some sections, and Ithaca College for released time and conference support. I appreciate the retreat centers that I have frequented for stimulating workshops and quiet spaces for reflection and work. Thanks to spiritual teachers and therapists who awakened and nourished insights that helped me see the world in a fresh way.

The translations are by the author, with the consultation of specialists. The use of the word "soul," reintroduced into western dialogue as a psychological term by James Hillman, influenced both Thomas Moore and me. Thanks to Hillman for his inspiration. Readers interested in exploring the biography of Carl Jung are referred to Dierdre Bair's biography (see References). Background on Heidegger can be found in Zimmerman and compared to Von Braun in Michel, Piszkiewicz, and Stuhlinger.

Finally, I want to stress that I have the greatest respect for the many conscientious, serious scientists, engineers, industrial designers, and other concerned makers and users of technology. Their methods of detached observation, collective verification, and so forth, are obviously valuable. I seek in this book to re-think the deep meaning of some unconscious enchantments about technology and to urge all to take responsibility for their effects. I know that many involved in the technological endeavor are concerned about the same problems that I am and seek solutions as urgently. May these words take wing.

1 *Unthinkable Enchantments*

Enchantment

Enchantment is a fascinating spell that takes over consciousness, a state of feeling that immerses the soul in dreamy reverie or fearful anxiety. You walk into a sun-drenched field in a forest and feel the wonder of that magic circle. You stumble upon a glorious waterfall, gaze at the immensely distant stars, float naked in a hot springs. You fall in love and are captivated by your lover's charms. You are possessed by a vision of a successful career, a brilliant invention, a dazzling work of art, a passionate determination. You dance with the Torah in Jerusalem. You kneel reverently for the *Ave Maria* in a gothic cathedral. You stand for the majestic "Hallelujah" chorus of Handel's *Messiah*. You bow in submission to Allah, or sit meditating in peaceful bliss. Enchantment brings life-enhancing ecstasy, and the soul demands it, thrusting us into reverence, adventure, dance, drink, or rebellion. The heart cannot survive without enchantment's nourishment. Quests to feast at that banquet drive us to passionate impulses, including fascinations with technologies.

Like a wizard's spell in a folktale, enchantment can also be destructive, seizing us with murderous jealousy, furious rage, deep disappointment, defensive fear, foolish denial, or exaggerated suspicion. It can cloud judgment and wreak havoc. You want to grab a gun or start a war. Industrial societies have the advanced intelligence to create nuclear submarines deep in the ocean full of missiles and robotic missions to Mars, three hundred million miles away, but stubbornly refuse to take serious responsibility for the spreading disasters inherent in industrial pollution. We have the information to solve problems from overfishing to ozone depletion, but we lack the will to act earnestly on that knowledge. Some groups even actively fight to continue damaging the planet.

When Greenpeace publicized the illegal logging of mahogany in the Brazilian Amazon forest that was being sold in the United States, they were charged with crimes by the U.S. Justice Department (Pfarr 2–6). Reason is insufficient to stop such absurdity. We are possessed of an incredible array of enchantments, "invisible walls" (Seidel) that can block action on rational goals.

Industrial society calls these enchantments "subjective emotions" that must be corrected by "objective reason" alone, which is generally seen as the sole arbiter of truth. But the purpose and vitality of the soul's enchantments are like a sweet nectar we cannot resist. So the negativity must be tamed by rational judgment and refined enchantment. The refinement of enchantments is the work of thinking, therapy, and spirituality, deepening our sense of wonder about existence itself, respecting our planet's life-system, and changing our purposes and personalities. An adequate understanding of enchantment must incorporate but go beyond pure reason, beyond sociology and politics, into a spiritual exploration of the soul. This is especially true of analyses of obsessions with technology, where the deeper soul has been too often ignored.

Enchantment is not a narrow, dysfunctional, trancelike stupor, or a bleary daze. Enchantments are common, ever-present factors of conscious-ness, whether mild or strong, denied or obvious, positive or negative. Music, drumming, dance, robots, rockets, movies, and drugs can produce "altered states" that can expand consciousness, but the "ordinary" consciousness of various cultural styles includes socialized enchantments that refine conscious-ness, from table manners to sexual ethics. Religions are naturally part of this cultural system, pointing beyond ordinary consciousness. The ancient Vedic Hindus, for examples, believed the greatest enchantment of all to be the *mahamaya,* the Great Illusion, the apparent reality of earthly life. Judaism and Christianity's creation text sanctifies high ethical standards, but also the human ecological enchantment of dominion over nature, and the social illu-sion of male domination of women. Some enchantments are pathologically abusive, some are taken for granted, and some, such as meditation or prayer, can enhance creativity, deeper understanding, transcendent awakening, and faith in the midst of crises such as death.

We can distinguish between "trance" and "enchantment" by their scope. A trance is typically seen as a more personal, intimate, hypnotic, or ecstatic state, detached from the environment, as in daydreaming or sleepwalking. Trances narrow the scope of consciousness, reducing one's scope of awareness. A trance may be an extended, extreme state, as in the "zombie" phenomenon, or a prolonged but changeable state, as in a neurosis or psychological complex that can bring suffering. A trance may also be a temporary, even self-induced state of relaxed focus in hypnotherapy. Trances may be pleasurable or pathological, positive or negative, addictive or liberating, and may appear in business or in romance. Negative trances may be treated with positive trances in therapeutic settings, as hypnotherapists such as Stephen Wolinsky and Dennis Wier describe.

By contrast, enchantments have larger, social settings, with major collective implications. They overlap with trances at times and can be negative or positive, mild or extreme, temporary or long-term. Enchantments may narrow the focus of a society's consciousness into a consensus agreement on certain beliefs and behaviors, such as optimism about technological progress. Enchantments introduce certain meanings into cultural life that take on a serious, rational tone but have a deep undercurrent of emotional and imaginative power. Devotions to hometowns and nations are enchantments, as well as passionate beliefs about the nature of reality, whether materialistic and random, cold and nihilistic, or spiritual and part of a universe created by a compassionate and just divinity. Enchantments are archetypal, living, vital experiences with deeply unconscious currents, such as myths and rituals about outer space or robots.

Enchantments are likely to be more strongly integrated with collective rational discourses. What may seem obviously rational to some, such as patriotism, utilitarianism, or quests for power and pleasure, have deeply unconscious enchantments rooted in desires rather than reason. Wars, national frontiers, and manic economic stampedes for wealth are examples. Enchanted desires may use rational tools to achieve their ends, thus giving projects a veneer of rationality while in the grip of a powerful, compelling enchantment. The meaning of "rationality," of course, gets mangled in the process, as in the debate over whether life can be created artificially. Even seemingly pure technical reasoning has underlying purposes, such as faith in technological progress or a godlike quest for ultimate power. Just below the surface, apparently "pure" rationality is in bed with enchantments.

The larger scope of enchantments extends beyond the clinical to the philosophical and religious because enchantments can expand consciousness significantly. Positive enchantments are cultivated by religious practices, although they can too easily be distorted by negative enchantments, such as nationalism. The finest religious faith involves a humble ego in the service of a higher being. Consciousness is expanded to set oneself in the larger context of a cosmic force. In Zen Buddhism, for example, the point of quiet meditation is not to withdraw the ego into a contracted, intimate trance but to open it to the vast scope of the Buddha-mind. As the hypnotherapist Stephen Wolinsky says:

> My *no-trance* state and Erickson's therapeutic trance are equivalent to the experience of *not* narrowing one's focus of attention by identifying and attaching.

> As mentioned, samadhi and other high states of meditation take this experience even further by dissolving the very boundary between subject and object. (46)

As mortal beings, we cannot be free of enchantments. There can be no non-enchantment, since all thought, even from the deepest sources, comes through earthly imaginative forms and the brain's desires, as basic as bodily survival and as refined as an elegant symphony. The goal is to free ourselves from the destructive and crude passions so we can flourish in the garden of the deepest, truest, most refined enchantments.

My emphasis here is on enchantments that are quite central to the collective consciousness of technological society, whether positive or destructive. They shape the development and use of technologies and feed faiths (such as progress) and utopian programs (such as human habitation in outer space) as well as dangers (nuclear radiation). They are ordinary and extraordinary, elegant and crude, inspiring and dangerous.

My use of the term "enchantment" in a sense broader than the conventional usage is intentional. First, it places this discussion in the tradition of the intellectual conversation about the "disenchantment" of the world by scientific thought and technological impact. Obviously there has been a major disenchantment of nature in industrial societies, but to understand it, we need to expand the use of the term "enchantment" to include a broader spectrum of phenomena, from private feelings to major paradigms. Only then will we realize the extent of the "disenchantment" and expand consciousness beyond the narrow range of industrial culture's conventions. This is not to invite a made-up "reenchantment," or a return to archaic superstitions, but to uncover the ageless, ever-present collective enchantments that have been suppressed and denied.

"Enchantment" here has a broader meaning than the psychoanalytic "unconscious." It includes both unconscious dynamics and those closer to consciousness, which are easily accessed. The word "enchantment" at root means to be "in-the-song," so one can imagine its intended meaning by joining in a song and feeling the deepening of cognitive functions by poetic, preverbal psychological, and spiritual dynamics. Try it with a national anthem, a favorite child's song, a classic such as Handel's *Messiah,* or "Amazing Grace." These are clues to its meaning.

Enchantments sneak up on us unawares in pre-reflective enthusiasms, and may even be combined with practical concerns. Some machines are primarily utilitarian (plows), while some are more enchanted (spaceships), but look deeper to see the premises, fascinations, and taken-for-granted assumptions and you are approaching the enchantments—speed, aggressiveness, and cosmic wonder. Many see technologies as tools to be controlled by human subjects standing outside them. Others stress that technologies are part of a wider horizon of culture, products of human thought and values. When we examine enchantments we go deeper still, into the unconscious depths that shape our motives, values, and decisions in the dark basement of the soul. Then we see that our machinery is not only a utilitarian necessity, or an autonomous realm of deterministic forces, but rather enchanted technologies designed to slake our endless thirst for speed, comfort, pleasure, power, and even transcendence.

This dimension of technology is widely ignored and suppressed as "subjective fantasy," but nevertheless enchantment chugs away at the heart of our machines. In principle this view is firmly opposed to the strict dogmas of modernism's quest for objectivity. Questioning dogmas and perceiving the enchantments is our task, so we can release the heart to its depths and open our eyes to the deeper imaginative, mythic realities of our technologies and related sciences. As the depth psychologist Thomas Moore says in *The Reenchantment of Everyday Life,*

> Enchanted science would not be devoted exclusively to facts, but it would be able to reflect back on its own mythic nature and its own fantasies and fictions. It would acknowledge that the very notion of facts is part of a mythological worldview. Taking itself literally and denying its place in the realm of imagination, science often closes the doors on enchantment. . . . A technology sensitive to enchantment would serve the soul. (45–46)

Our use of the word "soul" is an intentional break with industrial society's resistance to using it. It has meant many things over the years, especially the eternal soul embodied for a time on earth. The Greek word *psyche* literally means "butterfly," suggesting the soul's delicacy, autonomy, and departure from the dead body. Plato emphasized the importance of the eternal soul present amid the stars and reincarnated on earth. For Plato the purpose of philosophy is to recollect what that eternal soul knows. An ancient Hebrew term for soul is *ruach.* I use the term here to mean not only the eternal soul but first human psychological experiences, including unconscious as well as conscious feelings, moods, beliefs, metaphors, fantasies, and enchantments. Then I will explore a sense of soul in the world, *anima mundi,* or those enchanting relationships and phenomena where we often find our deeper selves, in a realm far larger than psychological subjectivity.

Similarly, we use the term "myth" not to mean "error" but to refer to a meaningful story, whether fictional or historical blended with fiction, to symbolize the soul's collective passions, such as myths of national origins and

conquests, romantic adventures, or heroic contests with evil. Myths are revealing expressions of collective unconscious enchantments, as Carl Jung and
Joseph Campbell show. Science fiction is the mythic text for technological
culture's fascination with unlimited power, heroic conquests of evil aliens,
endless technological brilliance, and even a quest for ultimate transcendent
truth. Technological enchantments are not limited to science fiction, however.
The philosopher Frederick Ferré says that we are "hypnotized by technique"
and that technologies are "incarnations" of our knowledge and values (135,
41). There is no technology without values, he stresses, and science cannot
function without ideals such as the perfect machine or the pure object (44,
87–89). Such enchantments inhabit the very machines we invent and use and
the rituals we enact with them, from robot games to human space travel. We
must recognize the positive and negative enchantments in our technologies.

Taking Out the Magic

Once upon a time the world was filled with enchantments, many of them
beautiful and harmonious, such as a wondrous sense of participating in nature
guided by invisible spirits. But some were destructive, such as the ancient fears
that drove tribes to bloody sacrifices. Religions were the homes of many of
these negative enchantments, so discrediting religion was a goal of scientific
and technological progress in order to clear the way for new foundations
for thought. We can be glad that many of these ancient enchantments have
been discredited and the horizons of knowledge have expanded. Today, many
fervently believe that we have removed all enchantments, but we have not.
They are simply different, now located in technologies, and they still have a
powerful effect on us, including the demand for sacrifice.

After millions of years of enchantment with the spirit nyads in rivers,
dragons in mountains, ancestors in the stars, and gods throwing thunderbolts,
it seems that humanity has, in a few remarkable centuries, shoved aside this
world of transcendent wonder. Industrial societies, it seems, have rapidly
replaced it with an awesome, immense surge in technological power held in
human hands, from new medicines and computers to nuclear weapons and
space travel. This apparent disenchantment was cultivated by many thinkers
and tinkerers, from the pre-Socratic philosophers on, and gained strength
by the time that Max Weber spoke of *Entzauberung*, or "taking the magic or
enchantment out" of nature:

> The fate of our times is characterized by rationalization and intellectualization
> and, above all, by the "disenchantment of the world." Precisely the ultimate

and most sublime values have retreated from public life either into the transcendental realm or mystic life or into the brotherliness of direct and personal human relations. (Weber 155)

Since the European Renaissance, the traditional history goes, one religious and mythic belief after another fell before the logic of humanistic, rational, and technological discoveries that reshaped the world from superstitious dogmatism into a cultural world of enlightened freedom and technological wonders. Copernicus's telescope opened up the heavens, Pasteur's lab revealed microbial sources of disease, and Edison's lab invented incandescent electric lights, phonographs, and cinema. It has become a commonplace belief that the modern world has been despiritualized by science and technology, that "enchantment's veil" has been withdrawn to reveal "cold material laws." As Carl Jung observed, "No voices now speak to man from stones, plants, and animals, nor does he speak to them believing they can hear" (Jung, "Approaching," 85). The modern world seems to have finally discarded erroneous emotional froth and replaced it with undeniable "realistic" factual reality, firmly fixed on the foundation of a subject/object metaphysics. Not so. This apparent disenchantment, I propose, is a strong surface phenomenon, and many valuable benefits have come out of it. But underneath surges a vast sea of unacknowledged, influential desires, passions, and quests for spirituality. For example, cars are not simply neutral transportation objects but beloved expressions of soul. People commonly speak passionately about "image" in cars:

> From the teen dreaming about a hot set of wheels to the self-imagined sophisticate, it is image that dictates our purchase. . . . Most of us can't imagine why anyone would buy a Hummer except to flaunt his financial ability to conspicuously consume. . . . Anyone who doubts the role of image needs only drive a rust bucket. (Burr 6B)

Incomplete and Lopsided

For the most part, this disenchantment has been happily embraced for its beneficial technologies and expanded knowledge. Modern technology has brought many liberating forces into history, from the virtual elimination of a short and brutish life of backbreaking work and slavery to the increase in freedom from hunger and cold to improvements in education, birth control, transportation, and communication. For these we can be extremely grateful. But the destructive effects of technology have always been criticized, such as the nineteenth-century romantic image of Frankenstein's monster, the Luddites' rampage against machine-induced unemployment, horror at the military potentials for technological advances, notably nuclear radiation, and contemporary ecological activists' warnings.

Increased leisure time has not resulted in the hoped-for raising of cultural levels as much as the spread of the mass media's lowest common denominators. Some critics have bemoaned crass materialism, blaring commercial pressures, modern alienation from nature, and the sense of meaningless absurdity that the technological worldview entails. Others criticize the lack of ethical guidance to direct the juggernaut of progress. The devastating wars of the twentieth century showed that technological wonders can be used for horribly destructive ends. By the end of the twentieth century the ecological crisis, which was originally seen as an unexpected side-effect of technological power, came to be seen as an inherent part of its rapid use of resources and careless pollution. The nuclear radiation problem is perhaps the worst of these ecological problems.

In 1949, June Casey was a college student near the Hanford, Washington, Nuclear Reservation, the major factory for making U.S. weapons-grade plutonium. The plant made an intentional, secret release of 5,500 curies of radioactive iodine—a multiple of thousands beyond the fifteen curies released at the later Three Mile Island incident. June's thick brown hair all fell out and never grew back. She developed extreme hypothyroidism, chronic fatigue, and had a miscarriage and a stillbirth, not understanding why. Thirty-seven years later, in 1986, she saw a small newspaper article that for the first time disclosed the release of radiation. It was a "planned experiment" (Glendinning 43–44).

The ecology crisis is only one example of the way disenchanted technological society neglects crucial ethics and purposes while claiming to be neutral. While technology has helped relieve many past social injustices, such as slavery, it has not eradicated others, such as the oppression of women, the removal of resources from poor countries, the accumulation of wealth at the top of "democratic" society, intentional abuses of power, the angry terrorist reaction to western culture's excesses, and deadly nuclear "planned experiments." These processes are not value-neutral but are solidly based on ethical decisions, positive or negative, conscious or unconscious.

There is something badly incomplete and lopsided about the industrial revolution that is ignored in the excitement about its achievements. Beneath the surface excitement of space travel and cell phones, there is an ethical hollowness, a terrible distress and anxiety, rooted perhaps in feelings of helpless victimization. Even the philosopher Bertrand Russell, who sympathetically chronicled the rise of modernity in *A History of Western Philosophy*, expressed some doubts:

> Technique conferred a sense of power: man is now much less at the mercy of his environment than he was in former times. . . . Unlike religion, it is ethically neutral: it assures men that they can perform wonders, but does not tell them what wonders to perform. In this way it is incomplete. In practice, the purposes to which scientific skill will be devoted depend largely on chance. . . . Ends are no longer considered; only the skillfulness of the process is valued. This also is a form of madness. (Russell 494)

Madness? Even a sympathizer recognizes a deep disturbance in the heart of industrialism that forcibly ignores its problems in the rush for riches and power. This madness, sometimes called "nihilism," is a terrifying prospect for some but is largely denied by others, leading to major political conflicts and crises, such as the toppling of New York's World Trade Center in 2001. Since double-edged technologies can be used for good or evil, denial of the unthinkable destructiveness of technologies allows blind continued production of potential disasters. But for many, even discussion of such problems is taboo.

In the face of such dilemmas, we need to step back and examine some foundational ethical, political, philosophical, and spiritual presuppositions. This book questions one of the central tenets of the technological worldview: Has the world really been despiritualized? Has nature truly become disenchanted? Is the world actually nothing but a set of dead objects inhabited by isolated subjects? Has industrial society really succeeded in denying all the purposes of life and machinery, turning it all over to chance and logic, thereby rejecting all the spiritual and ethical depths of humanity and the meaningfulness of the cosmos? Or is the disenchantment of the world only one powerful worldview, a surface phenomenon?

Worldviews guide our vision to make us see only what fits into that picture, ignoring or denigrating what does not fit. Technology's worldview elevates the importance of machinery and devalues the ethical and spiritual principles that question it. The disenchantment of the world into soulless objects in search of "neutrality" is basic to this effort to shove aside distracting principles. But this disenchantment has become a conservative defense of some old Enlightenment principles that are defended by its adherents who praise its benefits but tend to neglect its problems, due to their enchantment with its worldview. Langdon Winner calls industrial society's adherents "technological somnambulists, wandering through an extended dream" (Winner, *Whale,* 169). If we look below the surface of this claim of despiritualization, we will find what has been swept under the rug by the sleepwalkers, and then we can see that technology's world is not as rational as it claims to be.

It is foolish to champion a return to a premodern world that rejects the best of technology and science. This book is not "antitechnology," "dystopian," or "Luddite." Those who throw around terms like these I call "Titanic-ites." After millennia of cold, hunger, and disease, many technological benefits are obviously welcome. And many of the old prescientific beliefs of the past are best left behind, such as naïve, literal interpretations of religious scriptures as if they were purely factual, or what we call "history," which is full of mythic themes, such as Manifest Destiny. The problem is that the destructive effects of technology, from weapons to ecological crises, are not simply external side-effects that can be controlled with more technology. They are built-in problems. We must examine the depths of industrial society's presuppositions that allow us to stumble into the darkness, blind to our own destructive motivations. To do this we must examine both the psychospiritual level and the level of worldview, metaphysics, and ontology.

Worldview, Metaphysics, and Ontology

We typically use the term "worldview" to mean a collective frame of reference that is necessary to organize scattered perceptions into a coherent picture of reality. For example, a strong materialist worldview proposes that everything can be explained by scientific method, that matter is the basis of all reality, that reality is fundamentally mechanical, and that technology is morally neutral. It is impossible to think without organizing the world's chaos into systems such as this. For instance, in a sunset is the sun moving, or the earth? In the ancient Ptolemaic worldview, the sun moved over a static, disc-shaped earth. In the post-Copernican worldview, the earth circles the sun. Which is a fact? Superficially, the same sensory experience can be interpreted in opposite ways, according to one's worldview. Worldviews have been changed by technologies such as the telescope, which Galileo used to shift the West out of its old geocentric worldview. As Thomas S. Kuhn points out, worldviews are comprised of paradigms, or models for systems of knowledge and research, such as Newtonian physics or holistic medicine. They combine to make worldviews, coherent and comprehensive or not.

But the concept of a "worldview" retains a subjective sense of looking out at the external, objective world with a point of view. Thus even the word itself is not neutral, for it conveys a dualist metaphysics of a detached subject viewing a world. Philosophers and theologians have historically attempted to determine the nature of ultimate reality in "metaphysics" (beyond physics).

By "metaphysics" we mean the traditional rational analysis of the logically necessary and universal aspects of ultimate reality, such as Plato's forms, Aristotle's First Mover, or Hegel's Spirit. Medieval Christian metaphysics became highly speculative and stimulated a reaction leading to more rational and empirical metaphysics. Metaphysics has been characterized since Descartes as clear and distinct ideas supported by the implicit certainty of mathematical proofs. Skeptical empiricists, such as Hume, whose metaphysics trusted sensory data more than speculations about invisible essences, rejected metaphysics as needless speculation, usually involving an invisible God. A central premise of the metaphysics of science and technology is the Cartesian subject/object dichotomy.

Modern existentialist and phenomenological thinkers have agreed with the hazards of deductive speculations claiming dubious certainty and have responded by examining the global structures of the ordinary experiences of ultimate reality with a more openly intuitive methodology, changing the name to "ontology." This word is based on the Greek word *on* for Being, or the very "is-ness" of existence itself. Some think of Being as ultimate reality. Martin Heidegger made the major contemporary contribution to ontology with *Being and Time* (1927), in which he explored *Dasein,* or Being-in-the-world. He built on the tradition of phenomenologists, who find that intuitive common sense and down-to-earth experience uncover the more primordial reality of Being than reason alone, since it is impossible to limit experience to logical certainty. These latter assumptions are most compelling, because metaphysics can be excessively rationalistic and limited by speculative logic. So this book will seek to dig down to the ontological foundations of technological culture. Even though any metaphysics has intuitive, imaginative presuppositions and thus is not as logically certain as many have thought it was, I will still use the term "metaphysics" at times to name the thinking about ultimate reality that relies more on conscious logic than the enchanted depths of thought.

Before the ascendance of the subject/object metaphysical premise in the eighteenth century, God had to be pushed into the remote heavens by metaphysics, so as to leave the world free to become objects for human domination. Concomitantly, the breach between reason and faith grew wider. As this ontological split developed historically, the dualist gap deepened. Marcel Gauchet, in *The Disenchantment of the World,* writes: "[T]ranscendence not only separates reason and faith, it also divides subject and object" (53). As God was pushed further and further away, the subject/object split became wider: "The distinction between the bearer of knowledge and the object phenomena cannot be made without the deity's complete withdrawal from the world" (53).

Despite the relegation of God to the remote heavens by philosophers and theologians promoting the industrial worldview of a dead world, a historical thread of religion persists. Mysticism keeps the flame of spirituality burning. The ineffable *experience* of communion with the divine, beyond conventional horizons, repeatedly renews religious forms. Hildegard of Bingen, for example, said, "'Everything lives in God, and hence nothing can truly die, since God is life itself'" (34). But for many the subtle transcendent insights of mystics are outweighed by the industrial worldview of the emotional, subjective ego versus rational, objective technological mastery. In this framework, modern religion is increasingly reduced to merely a matter of personal, private, subjective interiority. The subject gains autonomy from external spiritual authority and develops political liberties and rationality. The object-world becomes an increasingly alien, dead, estranged, and neutral space and time field for exploitation to serve the subject. Questioning this metaphysic is unthinkable but necessary to those living within its horizons. The horizons of cognitive, rational thought must be pressed farther out in order to solve the urgent problems that its technologies have developed. Serious problems cannot be solved within the framework that created them, because the premises remain unquestioned. Some of these premises are expressed in various theories of technology.

Determinist Theory

A dominant theory of technology in industrial society is determinism, the view that humans are essentially *homo faber*, or tool-making animals, and that technologies are value-neutral tools, products of inevitable scientific progress. This theory is often part of a worldview that sees humans as part of a natural world that is indifferent and even hostile to human needs. Technology is thus seen as humanity's liberating power. Trains were technically superior and thus replaced stagecoaches and canals inevitably, in this view. The social consequences, such as the rise of new towns on the train line, were a direct consequence determined by the new technology. In determinist theory, technology and science are seen as nonpartisan tools, objects in a physical world that is impartial to their social uses. Autonomous technology has its own trajectory of efficient development, and humans must adjust to its imperatives.

As weapons advocates say, assuming value-neutrality, "guns don't kill, people do," which conveniently releases gun manufacturers and politicians from ethical responsibility. As pure tools, technologies are judged by the rational norm of efficiency, not the consequences they may have. If a machine is meant to be destructive and performs well, it is judged to be a good machine, efficient at its task, no matter how much damage it may cause. But the measuring stick of "efficiency" begs the question: Efficient toward what goals? Efficiency at blowing up people is obviously different from efficiency at healing the sick. Pure determinism is theoretically indifferent to politics and ethics, sociology and psychology. It is concerned with the rational, causal aspects of reality, and any other factors are conveniently dismissed as "external."

Determinism sees social change as largely driven by technologies and considers ethical or environmental impacts not as integral to the machines but as external "trade-offs" or "side-effects." So environmental and social consequences are largely ignored. Langdon Winner shows determinism's neglect of the political qualities of technologies in machines built for domination. Determinism has allowed the corporate and military dumping of technological poisons, such as "depleted" uranium, into the environment for the public sector to clean up. It also allows widespread media glorification of technological violence with the assumption that moral responsibility remains primarily in the hands of the users, not the makers, of violent films or video games. This narrow, technically focused view is blindly inadequate to consideration of the broad scope of technology's realm. There can be no technological imperative independent of human motives.

Instrumental Theory

One response taken against the determinist perspective is instrumental theory, the view that humans are far more than merely tool-making animals because culture—language, symbolism, thinking, and beliefs—is more fundamental to human nature. Organic processes far outweigh mechanical ones. Lewis Mumford said that "it was through the social activities of ritual and language, rather than through command of tools alone, that early man flourished" (Mumford, "Human" 63). Thus technologies are not value-neutral, blunt physical forces that push people around; they are secondary tools whose use is guided by the primary values of their users. Train track routes were not simply engineering decisions but value decisions. Some towns with successful canals rejected a train station, hoping the train would not replace canals, so the train took commerce elsewhere, and canal towns died. This social choice shaped the technological instrument. Police are carefully trained not to shoot their guns except under carefully defined rules of engagement, such as when a suspect points a gun at a person. This guideline requires the professional, trained user to value human rights. The development of the personal computer came not simply as a result of the new microchip technology but from the social milieu of the California counterculture that guided the development of Apple Macintosh personal computers, expanding their uses beyond offices.

Another important aspect of instrumentalism concerns beliefs about the relationship of science to technology. Some see science as the origin of technologies, but many medieval tools, such as the water wheel, were developed by craftsmen with little "scientific" knowledge. Now science and technology are a well-integrated system, heavily dependent upon each other, as with microscopes, telescopes, or computers. Science needs the data provided by technologies, as basic as mathematical techniques, and technologies need scientific principles to be built and operated. Each one can be conceptually distinguished, but in practice they are intertwined. Science is embedded in its instrumentation. Scientists cannot distance themselves as if they were merely theoreticians free of responsibility for use of the "applied" technologies, such as nuclear energy, that they create or use. Creators and users of technologies must exercise wider responsibility and control than determinism requires.

Instrumentalism is optimistic about the ability to control technologies. Nuclear engineers express confidence that their systems will be safe, but this optimism has increasingly been exposed as a faulty assumption when "unexpected side-effects" explode in our faces. Children do get guns and shoot their classmates. Computer hackers do cause havoc. Nuclear radiation does leak or explode, making regions around Hanford, Washington, and Chernobyl, Ukraine, into cancerous hot spots. The depth of human interaction with machines, from anger and rebellion to pride and recklessness, pries the controlling hand off the machines too easily, as many science-fiction films emphasize. Instrumentalism is an inadequate theory that is too optimistic, even utopian, to prevent technology from getting out of control.

Sociological Theory

Sociological theory of technology, the view that technology is not simply a set of tools that shape society or is subject to users' responsibility and control but is the product of a total social-political-economic culture, offers a broader methodology. Technologies are seen not only as physical machines. Broadly defined, technologies are "techniques," as the French philosopher Jacques Ellul defines them—intellectual tools such as mathematics, surveying, military organization, and sailing as well as concrete machines such as computers. Ellul places technology within a sociological context, but the deterministic power of technology made him pessimistic about the possibility of controlling its negative effects. Regardless of ideology, he said, capitalist or communist, technique has a powerful and autonomous effect, becoming the defining characteristic of industrial societies. The apparent autonomy of technology seems to have an unstoppable impetus of its own. After World War II and Hiroshima, Ellul was alarmed that scientific technique had become autonomous and out of control: "[W]hat seems most disquieting is that the character of technique renders it *independent of man himself.* . . . There is therefore nothing of sociological character available to restrain technique, because everything in society is its servant" (Ellul 306).

When he saw technologies having a dominant effect on society, Ellul felt rather defeatist, but he had a somewhat narrow view of sociology's analytical power: "The reality is that man no longer has any means with which to subjugate technique, which is not an intellectual, or even, as some would have it, a spiritual phenomenon. It is above all a sociological phenomenon" (306). If we do not go deeper into the roots of the problem, the surface autonomy of technology and its rigid rationality remain a discouraging, dangerous threat.

A fresh sociological view is taken by the "social construction of technology" movement, illustrated by thinkers such as Trevor Pinch and Donna Haraway. They show with important interpretive flexibility that technology is not simply the result of a bedrock of scientific facts or user controls alone. Broad social forces and actors in technosocial networks clearly shape and even construct technological development, from bicycles to rockets, thus demonstrating that technology is not as autonomous or dystopian as Ellul feared. For example, the informational "fog" of war is notoriously influenced by the interests of nations and militaries as they to a great extent "construct" rather than simply "discover" facts (Collins and Pinch 28–29). Whatever world is "out there" is difficult to clearly define, for researchers cannot be purely objective observers. They are fully involved participants in the social networks of research and production of technologies. Feminists have shown how gender bias is strongly involved in technology, such as the "smart" computer-controlled house, called by feminist critics a "deeply masculine" version of a house (Berg, "Gendered" 301–13). Machines that may seem technically "best" for one group (employers or men) may not be so for others (workers or women). Awareness of these social forces places more control and ethical responsibility in the hands of technology's creators and users. Sociological theory is essential to a full analysis of technology, and sociology's "social science" methods actually overlap into humanities methods, to their credit. A phenomenological philosophy is a fruitful ally to the social analysis of technology.

Phenomenological Theory

The phenomenological method opens broader philosophical horizons about the very nature of reality. This philosophy has many interpretations, so a selection of useful themes is needed. Edmund Husserl originated many important aspects of phenomenology, such as the rejection of scientific reductionism ("nothing-but"), psychologism ("only in the mind"), and scientism (science as the arbiter of all truth). For him, phenomenology intuitively brings to light the stream of experience in the lifeworld (*Lebenswelt*), far larger than the theoretically geometric world of subject-object metaphysics.

Husserl's successors, including Jean-Paul Sartre and Maurice Merleau-Ponty, developed his thought in various directions, but Martin Heidegger provides the most penetrating analysis of the nature of technological culture. Heidegger generated numerous ideas that stimulated important twentieth-century thought. Technological culture is transforming the entire world, he says, into a desolate realm of raw materials to be used only for industrial purposes. Seen through the metaphysics of the subject/object dichotomy, modern consciousness imagines itself detached from a world of objects that it can control, but this degrades even humans to mere objects. With no larger context, technological culture promotes a dangerous pursuit of the will to power, resulting in a nihilistic world, full of powerful and threatening machines.

Subsequently, Don Ihde has expanded technology studies with a phenomenological analysis. Technologies are not isolated objects, he stresses, because finally there is no "inside" or "outside." We are embedded in the lifeworld, not as victims of an autonomous technology but as codeterminants of our mutual fate. Our perceptions are contained within our hermeneutics (interpretive principles), and these we can change. Machines are by no means neutral; they guide perceptions, as with eyeglasses. He shows how metaphors such as the world as a giant clock become metaphysics and how such technological instruments guide scientific theory. Phenomenology opens the most important door to a philosophical excavation of many taken-for-granted concepts in technology studies, such as the subject/object worldview. Approaching technologies not as subjects standing opposite objects but as parts of the field of Being-in-the-world frees consciousness from the mental iron cage built by the quest for objectivity. Deterministic theory and instrumental theory are inadequate responses to the radical challenges of technological culture. Sociological theory and phenomenological theory have expanded the horizons of technoanalysis significantly, but further exploration is needed into the depths of collective soul and our foundational assumptions about reality.

Enchantment Theory

Strictly rational, analytical thought built on the metaphysics of a disenchanted world that has succeeded in building technologies and sees them as determinative is not the best mode of consciousness for understanding its own deeper cultural context. Technical consciousness cannot see outside of its own self-imposed boundaries, so it is incapable of solving its deepest problems. Reason is necessary but inadequate, for logic and technological fixes are insufficient. An entirely new consciousness is needed. Unconcealing the deep enchantments thinly veiled in conventional theories requires a broadening of the horizons of thought. The first step is to look into the deeper nature of technological thinking itself. There we see that it contains a far larger and more influential component of enchantment by unconscious fantasies than is commonly acknowledged. Imagination is not to be brushed aside as subjective, romantic sentiment. Rather, enchantments necessarily pervade and feed the heart of technological culture. There is no technical thought without enchantment because technological culture is teeming with dreams, visions, hopes, goals, expectations, and imaginative premises. Awakening to this realm will require a radical shift in consciousness already foreshadowed by sociological and phenomenological thinkers. Taking responsibility for these dreams frees us from the compulsive inevitability rooted in unconscious drives that so distressed Ellul.

For example, the problem of "technology out of control" is illustrated by those who argue that advanced research must continue unrestrained or we will lose technological leadership and "miss a revolution" with important economic effects (Clark). However, leadership, being on the technological frontier, and economic benefits are *not* neutral, rational reasons for research drawn from an objective, disenchanted world. They are very much part of the human, imaginative social-political world that easily pushes things out of control. Also, the unexamined premise that the world has been despiritualized in the discourse of science and technology is not a fait accompli but rather a rhetorical assumption, a narrow way of seeing the world.

As several thinkers perceive, these types of unexamined, enchanted claims have led us into a "technologically intoxicated zone" that is "spiritually empty, dissatisfying, and dangerous" because we turn our backs on its consequences (Naisbitt et al. 2–3). In this intoxicated zone is born the slippery bureaucratic saying about technological research out of control: "It is either too early to tell whether it works or too late to stop the program." The most resistant difficulty in undoing the enchanted dangers of the industrial order is not the physical nature of technologies but the psychological trances that motivate the machines (Berry 32). Unconscious psychological factors, such as the aggressive warrior spirit, must be considered if we are to understand "unexpected" consequences of technologies.

Sherry Turkle, a psychologist at MIT, has uncovered many important unconscious personal meanings of computers, such as the mindless, addictive "holding power" of aggressive, passionate, and eroticized games, childlike wonder, seeking feeling relationships, mastery, and seeing oneself as a machine, seemingly full of ego-boosting feelings of power and control, safety, and godlike mastery (Turkle, *Second*). Computers give meaning to lives, making users feel powerful, aggressive, in control, seductive, and even godlike.

Technology does not inhabit a neutral world of pure space, time, causation, and reason. Rather, technology's lifeworld is fully imbued with imagination, purpose, ethics, motivation, and meaning. The latter option seems absurd in industrial society's worldview, due to its foundational disenchanted subject/object dichotomy. But technology is not simply a set of "objects" in a neutral world distorted by "subjectivity." Technologies are *embodiments* of imaginative, collective desires as much as utilitarian, rational designs. Missions to Mars are rooted in the compulsion to know if there was ever life on Mars, which would impact modern cosmology and tell us we are not alone in the universe—not just fact, but fascination.

Some say that each culture, including ours, is indoctrinated with a kind of hypnosis, an unconscious enchantment that prevents its members from accepting the reality of "unthinkable" phenomena outside its worldview. This inability to see or accept phenomena that do not fit into the accepted framework led one well-indoctrinated skeptic to the absurd extreme of saying about this idea: "I wouldn't believe it if it *were* true!" (Harman 61). I will argue what to such people is "unthinkable": that the subject/object dichotomy itself is an enchanted, dangerously inadequate description of the foundation of the technological worldview, for it has all along been displaced by being constantly bridged and conjoined by a more primordial ontological union, like a wound that heals from below.

When we query some of the conventional, taken-for-granted, a priori assumptions of technological society, we discover that behind the bastion of reason lurks a zoo full of passions, desires, and enchantments, barely disguised by conventional consciousness. These unthinkable dynamics are numerous and beyond adequate description. They have endless names, such as obsessions, fascinations, crazes, fetishes, fixations, phobias, and so forth. I will highlight one term—"enchantments"—to capture briefly the vibrant, captivating power of these behind-the-scenes operators. Like characters in an ancient myth or fairy tale, these forces have many masks that slyly guide the plot, like the old woman in one version of "Beauty and the Beast" who enchants the rude prince into a beast. Even some physicists, such as Arnold Pacey, are saying that "technology-practice . . . is not value-free and politically neutral, as some people say it should be" (Pacey, *Culture,* 5). Pacey also declares the explicitly evil nature of some technology that, incredibly, is denied (*Meaning* 171–98). Objectivity's detachment is a useful but increasingly destructive practice.

Our machines are products of numerous unthinkable enchantments that use reason to fill passionate needs. The disenchanted, neutral world itself is not at base a factual reality or even a rational principle but a highly imaginative value, a mythic image promoted by industrial culture. The blend of savagery and technology in war should convince us of the presence of passions in technology. The continuing threats of highly dangerous technologies such as nuclear waste and biological weapons, which are badly guarded, underfunded, and subject to normal human fallibility, makes the acknowledgment of the inextricable human role in technology an urgent matter. We can easily construct a continuum between highly utilitarian technologies such as chain saws and highly imaginative technologies such as android robots. But digging deeper, we can even find imaginative elements in utilitarian tools, such as the phallic symbolism of chain saws.

Technological Desires

The Enchantments of Technology is about the fascinations, charms, captivations, mystiques, trances, wizardry, sorcery, and magic of technology. Technological wonder is the unconscious underside of our ordinary attitudes toward machines. These mythical enchantments are ambivalent: on one side we welcome technology's generous benefits after millennia of suffering overcome by its diligent inventors, intelligent scientists, and creative engineers. Yet on the other side linger the unexpected problems and even horrors of the machines that our industrial society has devised. This book is about the ways we are entwined with these desires that drive our technologies: we love them, fear them, and yet feel trapped by their ambivalent mythic enchantments.

A "technological mystique" interprets technological failures in the instrumental mode as experimental steps in the inevitable progress of technology, so a critical view that certain dangerous technologies should be restrained is brushed aside by utopian optimism. This mystique assumes that, in the larger picture, continued experimentation will inevitably bring success. But this assumption is flawed, because technology is not completely guided by scientific, rational principles. Economic benefits are eagerly sought in boardrooms and books such as *Technology and the Wealth of Nations* (Rosenberg et al.). Huge military budgets dominate technological research. Failure and high cost may be swept aside as favored projects are pressed forward, such as the Stealth fighter plane, worth its weight in gold, while urgently needed health-care, education, and ecological energy-generation projects go woefully underfunded. The mystique lies in our souls' passions, fears, and dreams.

Optimism about inevitable success is also defective because numerous "megamistakes" have occurred, unintended consequences or evidence of unfulfilled utopian dreams, as with the sinking of the *Titanic*. Technologies always exceed their expectations, either positively or negatively, because mere reason cannot predict their eventual impact. Yet in spite of tragic failures, optimism trudges on with technological fixes and ignores or denies the negative potentials when technologies "bite back," as with nuclear radiation leakages. The problem has many sociological, political, and economic dimensions, but here we will inquire into the collective, unconscious, psychological, philosophical, and spiritual dimensions.

Positive dreams, fantasies, visions, or utopias *and* negative fears, defenses, or furies are all inescapable enchantments. Consciousness includes them in its roots; the effort to achieve "objectivity" has unsuccessfully attempted to eliminate them. Yet they linger, and we need to recognize and make ethical decisions about them. Some enchantments are crude, greedy, violent, and destructive, as we see in criminal behavior. Others are refined, elegant, beautiful, and guide our finest behavior, as in compassion, social justice, and spiritual peace.

Machines Bite Back

A nineteenth-century French naturalist brought gypsy moths to the United States, expecting to start a silk industry. The caterpillars escaped and multiplied, and they continue to ravage forests. Corporations computerized offices, expecting a "paperless office" and increased productivity, but offices are now using much more paper than ever. Every year, numerous huge ocean vessels sink, dumping thousands of tons of metals, food, building supplies, and fuel into the ocean. For example, the *Prestige* (an enchanted name) sank off the coast of Spain in 2002, dumping twenty million gallons of crude oil (Zera 69; Daly, "Oil").

The reckless spills continue. On November 22, 2004, a Greek oil tanker was docking in the Delaware River near Philadelphia. It hit a large underwater pipe or pump long ago dumped overboard and lost about 473,000 gallons of crude oil, of which cleanup crews will collect only 10 to 25 percent; 25 to 50 percent will sink to the bottom, destroying numerous fish, crabs, and birds on over fifty-five miles of the river (Spotila, "Tragic").

"Normal accidents" must be expected in complex systems tightly linked with little time for human intervention to prevent disasters. Thus concludes Charles Perrow in *Normal Accidents: Living with High-Risk Technologies.* Oil spills, ship collisions, airplane crashes, space rocket explosions, and radiation leaks must be expected in such complex and dangerous technological systems, he argues. Catastrophes can lead to improved safety, with serious focus on protective measures, as has been shown with mine safety and air traffic control. Yet accidents cannot be totally eliminated. This is to a great extent due to industrial managers and politicians who judge it too expensive to institute needed safety measures, such as double ship hulls.

What does this argument assume? *Human desires are always involved in technologies.* Drunk driving and speeding are the results of machines misused by human passions. Technologies are not neutral objects placed out in the world by engineers that will perform flawlessly, independent of human interaction. The central expectation of control, the desire for mastery, is an integral part of technological intentions, but it is an inflated expectation—an enchantment—that can lead to terrible consequences. Either the original intention or the user's actions will inevitably have unintended side-effects. Initial expectations sometimes neglect critical factors. The term "side-effect" also indicates that the builders' intentions do not take into account factors that were considered peripheral and unimportant but which end up being disastrously central. The Iroquois Theatre in Chicago was imagined to be so fireproof that it opened in 1903 before its sprinkler system was operating. With no firefighting equipment, it burned down a few months later, killing over six hundred people (Tenner 19).

Early steam engines in ships and trains would explode if the engineer pushed them beyond their limits, as some did in races. Automatic steam-pressure relief valves were then required by government safety laws. Traffic jams persist in cities with more and bigger highways because the number of cars and trucks has multiplied. Edward Tenner documents this problem in *Why Things Bite Back: Technology and the Revenge of Unintended Consequences.* Who anticipated angry computer hackers, or grade-school children shooting each other? Computers and guns are not simply objects in the world that will behave as expected, nor does the industrial worldview take account of the entire human soul's dynamics. Obviously we must vigilantly institute safety measures, which reduce but do not stop disasters.

History shows that technological perfection is a utopian dream. Nevertheless, the dreams continue, and they urgently need critical examination. Machines, no matter how sophisticated, are not autonomous objects that will obey their masters' intentions. They can magnify the potential for harm. Instead of saying, "guns don't kill, people do," we should say, "with guns people can kill many more people, much more easily." Powerful technologies amplify human power, for better and for worse. We err dangerously when we neglect the human desires that are inherent parts of technological design and usage. When the dreams, expectations, fantasies, and passions that drive technological design and use are ignored, machines can cause massive, fatal damage. Industrial society's dreams of utopian technological accomplishments are fascinating. But they are mythic fantasies that need to see the light of critical consciousness.

Or, like a crocodile lurking under a swamp, technology's enchantments will leap up and bite back.

Biosphere: Fire and Air

The power of enchantment and the inevitability of unintended side-effects are illustrated by two efforts to construct prototypical outer-space human habitats. On an island park on Ile Sainte-Hélène, Montreal, stands a large spherical monument surviving the 1967 World's Fair. This giant ball showed off Buckminster Fuller's brilliant design of the geodesic dome, which was widely copied afterward. This dome still stands 250 feet in diameter; it was originally the United States exhibit for Expo '67. It displayed the pride of U.S. space technology, the frontier of industrial society's inventiveness and heroic quest to conquer space (La Biosphere).

But in 1976, during repairs, a welder's torch splashed sparks that set afire the clear acrylic skin. In a raging fury, the entire plastic covering sent flames high into the night sky, melting the plastic, leaving a naked steel framework. With the fire came crashing down the easy dreams of inhabiting other planets in this type of structure. Imagine what would happen if a similar accident occurred in outer space, how quickly devastating it would be to its inhabitants! In response to the growing environmental movement, the dome's purpose shifted, and by 1995 it had been transformed into an environmental educational center. Now it is filled with warnings about dangers that the industrial dream is causing to the earth's water.

Visit the upper-deck restaurant, look up into the beautiful structure, and feel the wind blow through the uncovered dome, like a spirit-wind blowing away a fading dream of space travel. Listen for the whisper of Bucky Fuller's dream: "I see God in the instruments and the mechanisms that work reliably, more reliably than the limited sensory departments of the human mechanism" (Fuller 2). Fuller's vision of God in technology's perfection is a common enchantment, not a rational analysis. It reminds us of the overblown, dangerous, and expensive captivation with the fantasies of the conquest of vast galaxies and, by contrast, the earth's delicate environment—home for our species.

A second, more ambitious project was Biosphere 2. Near Tucson rises a fascinating community of buildings whose design echoes Fuller's geodesic construction methods. Shimmering in the desert heat, a large rectangular greenhouse and human living space rises, seeming like a grand Mayan-looking pyramid of steel and glass. This is Biosphere 2, originally an experimental effort to build a self-contained, self-sustaining environment to explore the bold futuristic theory that a patch of the earth could be replicated on another planet to support space travelers. It is a $150–million, 3.15–acre terrarium containing its own desert, rain forest, oceans, and thousands of plants and animals. It was intended to be an air-tight home for eight volunteers who lived there from 1991 to 1993, independent of the earth's atmosphere, farming and feeding themselves: "Sealed off from the rest of the world, this space colony on earth will be home to four men and four women for the next two years" (Stover 54).

But problems began early. The pollinating insects rapidly died off, vines multiplied wildly, crops failed, nineteen of the twenty-five small-animal species became extinct, and trees shot up, then fell over for lack of wind to strengthen their trunks. Ants, cockroaches, and katydids got in and multiplied. The experiment failed when the interior oxygen was reduced to dangerous levels by unexpected factors such as the rich compost and curing cement foundations, which absorbed oxygen, lowering the oxygen level from a healthy 21 percent to a dangerous 14 percent. The experiment finally lost scientific credibility when the experimenters had to pump in extra oxygen to survive, as early as three months after they began, and again in February 1993 (Recer; "Oxygen" 4B; Maxson 56–60; "Biosphere Blues" 26). After eighteen months, the "Biospherans" had to exit, six months early. They had become hungry, short of breath, irritable, confused, and tired from oxygen insufficiency and constant work, and they had each lost an average of about twenty-five pounds. The delicacy of our earth's natural processes became evident, and the difficulty of reproducing them on another planet revealed how bold and enchanted the original vision had been.

These biospheres were built by mythic enchantments. Critics dubbed Biosphere 2 a "technoscience soap opera" and "a monument to scientific hubris" (Luke 157, 162). The New York Times Magazine recognized the myth of Noah's ark in the project ("Noah's Ark"). Today's version imagines that the world is fated to undergo catastrophic destruction, and human space travel is the only salvation for the few believers. Such utopian and end-of-the-world dreams inhabit these biospheres as well as the obsession with space flight and the pursuit of the latest frontier. These biospheres seem like two log cabins on the frontier of space, but space is a vast leap with far greater survival hazards than western prairies. The extravagant reach for outer-space conquests, combined with the manipulation of natural environments, implies the Promethean hubris of theft of power from the gods. "'You have to play God, and that's not easy,'" said one Biosphere scientist (qtd. in Appenzeller 1370). In 1995 the beautiful building in Arizona was temporarily converted into an ecological education center (Wolfgang). Like Montreal's Biosphere, Biosphere 2's space travel enchantment, literally and figuratively, was brought down to earth.

These problems have not deterred utopian human space-travel dreamers, however, who still envision futuristic habitations on other planets, apparently confident that technological solutions to the Biosphere problems will be found. The enchantment lives on, an aspiration undeterred by discouraging data. Such expensive technological monuments are visionary statements of technological virtuosity, not economical calculations. Peter Medawar, in *The Hope of Progress*, argues that a space probe, "like a cathedral . . . is economically pointless, a shocking waste of public money, but like a cathedral, it is also a symbol of aspiration towards higher things" (116).

Such technological monuments do not represent emotional froth, irrelevant to the practical rationality of technology. They show that the rational, materialistic goals of technoculture are fully entwined with the sense of adventure, exploring new frontiers, and the pursuit of "virtuosity for its own sake" (Pacey, *Culture*, 89). Pursuing virtuosity for its own sake means striving for technological expertise and mastery, not for the sake of practical, utilitarian purposes or even for rational purposes. It means following the star of pride and accomplishment, the fascination with discovery and mastery, the enchantment with technology as a monument to human greatness, much as the pyramids or cathedrals glorified a vision of divinity. These are not extraneous factors in technological culture; they are essential guiding myths, expressed in many a science-fiction book or film. These monuments to the dreams of industrial culture are purposeful, inspirational images. The challenge is to acknowledge them, instead of leaving them to operate unconsciously.

The philosopher Mary Midgely, in *Science as Salvation: A Modern Myth and Its Meaning*, stresses that the myths of technology serve not as "froth" but as important guiding stories:

> We have a choice of what myths, what visions we will use to help us understand the physical world. We do not have a choice of understanding it without using any myths or visions at all. Again, we have a real choice between becoming aware of these myths and ignoring them. If we ignore them, we travel blindly inside myths and visions which are largely provided by other people. . . . Throw purpose out through the door and it seems to creep in up the drains and through the central heating. . . . Attending to the workings of the scientific imagination is not a soft option, and it is not mere gossip. . . . It plays a part in shaping the world-pictures that determine our standards of thought—the standards by which we judge what is possible. (13, 15)

The Enchantments of Technology will explore the soulful and spiritual depths that linger under our apparently objective world, such as the will to power, the lust for speed, and optimistic, utopian, heavenly dreams of progress. We must bring them to consciousness so we can avoid falling victim to their enchanting powers.

Technology and Unconscious Soul

Some people have trouble accepting these ideas because enchantments are largely unconscious, and unraveling them requires openness to the dynamics of unconscious soul that strictly rational thinking does not welcome. Openness to unconscious factors can reveal how technologies are not simply tools we use in a social context but rather instruments crafted to express our identities. Speeding cars, luxurious airplanes, android robots—these are not only social objects but images, expressions of who we are as an industrial culture. The biologist Robert Pollack says in *The Missing Moment: How the Unconscious Shapes Modern Science:*

> The conscious part of science is what most scientists would insist is all there is to science. . . . [But] the unconscious parts of the mind of science . . . emerge as fantasies and obsessions shared by scientists in these fields. . . . Science is the product of the unconscious sources of imagination and introspection as much as it is the product of a set of rules. . . . Science engages the whole mind of each scientist, both the conscious and the unconscious parts; fears, fantasies, dreams, and memories are as important to a scientist as any measurement or model. (58, 75)

Depth psychology shows such unconscious images to be very important.

The meaningfulness of unconscious psychic factors was shown by Sigmund Freud, who laid the foundations for a bold new theory of interpreting the nature of the psyche or soul. There was no concept of the unconscious until Descartes sharpened the focus of the western mind in his method, forming what was later to be called "ego" and "consciousness." But the bottomless imagination would not simply fade away, so the romantics, Nietzsche, and Freud formed the notion of a region of mind called the "unconscious," in contrast to a Cartesian sense of clear thought. L. L. Whyte, in *The Unconscious before Freud*, says that Descartes, "by his definition of mind as awareness, may be said to have provoked, as reaction, the European discovery of the unconscious mind" (86). In technology studies, acknowledging the unconscious shows that there is an indeterminate amount of unconscious content in any conscious thought, symbol, or technology. Wish fulfillments and sexuality are common Freudian themes, so a strictly rational analysis of our machines cannot be complete. Rockets, for example, have a wishful and phallic dimension tied to wishes for flight's thrills.

In late nineteenth- and early twentieth-century Vienna, a revolution in consciousness was ignited by Freud, who showed in medical language the meaningfulness and power of the unconscious mind in its ability to influence consciousness and psychosomatic ailments. A paralysis can be caused by unconscious traumas (Freud, "Some Points" I:157–72). Through the study of dreams, myths, and their symbols, he uncovered a vast realm of highly influential unconscious dynamics, notably the infantile Oedipus Complex.

The philosopher William Barrett shows the spreadng influence of these dynamics: "The anthropological sciences, and particularly modern depth psychology, have shown us that human reason is the long historical fabrication of a creature, man, whose psychic roots still extend downward into the primeval soil" (Barrett, *Irrational*, 37). Depth psychology has pulled aside what was once considered the immaculate, autonomous region of rational thought and uncovered the passions it conceals, which guide and shape culture and technology. Jacques Ellul, who was fascinated with the way radio and sports cars appeal to emotions, saw the implications of depth psychology for technology studies: "There is no such thing as purely objective information. . . . Every technique, and above all every human technique, makes a fundamental appeal to the unconscious" (Ellul 364, 403). This radical intrusion into "pure" reason has shoved aside the claim to certainty and completeness in rational discourse, because there is always an unconscious level below the surface shaping thought. Unconscious subliminal perceptions and symbolic meanings are constantly influencing conscious thought, such as in the phallic implications of the distinction between "hard" and "soft" data.

Freud gave deeper meaning to the category of the subject by interpreting the symbols of dreams as purposeful expressions of unconscious complexes, organized primarily by myths such as Oedipus and Narcissus. Sophocles's tragedy *Oedipus Rex* became Freud's mythic text for helping patients who were too attached to parents of the opposite sex to release their infantile bond and grow into their proper adult emotional life. The subject's irrational, biased desires now had a certain logical structure and a therapy. Freud named the subject's unconscious realm of raw desires the "id" and the rational department of the subject the "ego." This ego regulates relations between external and internal reality.

Attempting to persuade his fellow physicians, Freud clung to elements of his nineteenth-century metaphysics of the subject and to mechanistic physiology; he envisioned the psyche contained in the skull and its dynamics as "mechanisms." Defenses, such as regression, repression, projection, and introjection, were seen as mechanisms rather than "automatic organic psychic dynamics." Freud's categories of "subjectivity" and "mechanism" helped define his sense of reality and sought to return control to the patient's subjective ego. But despite his mechanistic metaphors, Freud opened the door to unconscious drives, independent of ego and its machinery. Despite early scorn in the 1920s and 1930s, therapeutic psychology has been invigorated by the effectiveness of Freudian psychoanalysis.

Archetypal Psychology

Out of psychoanalysis emerged archetypal psychology. Originated by the Swiss psychiatrist Carl Jung (1875–1961), it has been expanded and refined in several directions. It began by rejecting Freud's excessive emphasis on sexuality and critique of religion as illusion. It continued the practice of talk therapy but developed the new view of the unconscious psyche as having a collective layer deeper than Freud's personal psyche. Many images and messages from the unconscious were seen to be not only about personal childhood events but as aspects of the grand, ancient human drama. A dream about your father is not only about your personal father but the Great Father.

The theory of archetypal images developed as the organizing principles of the collective unconscious. Archetypes are seen as unconscious, collective, psychological instincts, positive and negative, not censoring Freudian distortions but genuine symbolic expressions of the deep psyche. The view of the collective psyche slowly moved psyche out of the narrowly subjective category of the mechanical worldview into a broader ontology. Archetypal psychology offers a most helpful way of understanding the depths of technological culture.

Archetypes are innumerable. They can be perceived not in themselves but only through images appearing in phenomena such as dreams, myths, behavior, moods, the body, the environment, and machines. An archetypal complex or constellation could easily generate a strong enchantment. A classic archetypal image is the shadow, which contains all the crude, negative, and fearful evil rejected by consciousness, symbolized by monsters, villains, enemies, the myth of the devil, or the destructive, out-of-control machine. Each archetypal pattern has its shadow as well as its positive forces, and each has its patterns of developmental growth. In technology, the genius, inventor, wizard, trickster, and magician are ancient archetypal images projected onto technological heroes, as in Daedalus, Edison, astronauts, or robots.

The goal of archetypal analysis or therapy is not perfection or adjusting to society but healing by balancing the psyche's conscious and unconscious dynamics. In industrial culture a major balance needed is relaxing the conscious ego's domination of psyche and allowing more subtle feelings, moods, dreams, and myths to bring up deep unconscious dynamics. This happens often when the ego is defeated or humbled by a life crisis such as a loved one's death. Then the goal is not only to strengthen the ego's pride but to help it to hear the deeper soul's subtle symbolic messages that can lead to balanced healing. An example in technomythology would be *Star Wars'* partial balance between very powerful high-tech machinery and the overall power of the mystical Force guiding the Jedi Knights, gathered at a culminating point with the visionary images of some of their dead amid the primitive Ewok villagers. The religious theme would also be the victory of the Force over the Dark Side through heroic struggle. In archetypal theory, the ego learns to serve the higher Self, the divine perceived in the psyche, regulating to some extent the various and competing dynamics of the many archetypal instincts.

Archetypal psychology has helped expand religious thought by showing that the higher divine Self can be symbolized in innumerable ways, not only by an old man in the clouds. The great cosmic mystery can be perceived in a goddess, a tree, an ocean, a crystal, a mountain, an animal, a sparkling pond, a star, or a magical robot. Thus myths are not someone else's erroneous beliefs but highly significant narratives of collective unconscious dynamics—fighting dragons, loving a prince, flying high above, conquering an opponent, fearing abuse, or being saved by a divinity. In technological culture, machines take on mythical qualities, and ancient cultural and religious images reappear in technological form, as in UFOs.

Archetypal psychology began to break the bonds of the mechanistic world-view after Jung's near-death experience and his studies in alchemical mysticism. Then James Hillman took it further, saying in *Re-visioning Psychology,* "Subject and object, man and Gods, I and Thou, are not apart and isolated each with a different sort of being, one living or real, the other dead or imaginary. The world and the Gods are dead or alive, according to the condition of our souls" (18). This analyst also opened up further the *anima mundi* (soul in the world) and pressed psychologists to see psyche collectively in the world, not only in the interior subject. Hillman perceives that the depression in industrial culture could well be a delayed reaction to our hostile domination of nature. Perhaps it is a mourning for the destruction of the world by industrialism (Hillman and Ventura 45).

Archetypal psychology has also been greatly expanded by feminist analysis. One example is Jean Shinola Bolen's *Goddesses in Every Woman.* Letting ancient Greek goddesses constitute a typology of the archetypal feminine soul, she describes, for example, Athena as the urban career woman, Aphrodite as the passionate lover, Artemis as the outdoors woman, Demeter as the earth mother, Persephone as the meditative mystic, and Hera as the powerful queen. The queen would seek powerful machines, the lover would concern herself with beauty enhancement and birth control, and the earth mother would nurture ecology and family. Another Jungian analyst, Sylvia Brinton Perera, wrote *Descent to the Goddess: A Way of Initiation for Women* about the Babylonian Ishtar's encounter with death and *Celtic Queen Maeve and Addiction: An Archetypal Perspective* about the Irish goddess whose myths help heal addictions. Jungian psychology has matured considerably into an analytical and archetypal path that is less "empirical" and more poetically phenomenological, less "subjective" and more soul-in-the-world (Brooke).

Psychotherapy is an archetypal initiation rite. It dethrones the ego, taking it on a journey into the unconscious to encounter its archetypal depths. Through therapy, a person develops a healthy sense of the depth of soul beneath daily life and becomes able to feel the tug of deep drives and desires and give them a place in conscious life, refined in interaction with the conscious ego rather than remaining blindly unconscious and crude. This deeper consciousness applies well to technology studies.

It certainly seems strange to think along the lines of deconstructing the subject's ego. It is hard to imagine that a thinker is not in control of his or her thoughts or an agent is not the subject of his or her action (Descombes 121). But this is exactly what archetypal psychology can lead to, because the locale of consciousness is not simply a conscious ego in control. All these activities listen to a deeper source of thinking, acting, and desiring, an archetypal well, a back-of-the-brain unconscious world larger than the conscious agency of a subject. Obviously, ego's rationality is essential to thought, but it can learn to listen to rather than stifle archetypal depths.

The subjective ego cannot adequately determine reality by itself, for it is never without influence from unconscious, archetypal forces. In this sense, the whole of a person's psyche, both conscious and unconscious, collective and personal, may be called "soul." The lover, warrior, inventor, and astronaut are all rooted in archaic imaginal stories—not even identities, for that is too subjective, but in dreamlike legends that arise with endless thematic variations. As a lover, a man recapitulates Tristan, Romeo, and Don Juan. Warriors echo patterns of Gilgamesh, Moses, and Napoleon. Inventors breathe in the spirits of Daedalus and Dr. Frankenstein. Astronauts float in the skies of magicians and angelic souls. None of us stands alone without this collective orchestra of soul. This background does not by any means dilute one's freedom or relieve one of ethical responsibility, but rather it enriches a person's freedom and grounds ethical decision making about technology.

The alternative to the metaphysics of the subject is not "intersubjectivity." This term retains the reification of the subject into an entity that now somehow has to connect with other subjects. The interior castle of the subject bars the way to full communion, so "intersubjectivity" is an inadequate halfway construct. For example, some see aesthetics as intersubjectivity, a mirror of objective nature, a personal expression of inner feeling, or a subjective preference regarding beauty. But all of these options retain a subjective metaphysics and restrict beauty to an interior feeling. This restriction of the subjective realm to interiority discredits phenomena that cannot be explained as objective facts. Some methods of study attempt to quantify and objectify these but succeed only weakly. For example, on a scale of one to ten, how beautiful is a sunset or a spaceship?

Out of his study of alchemy as symbolic, mystical processes came Jung's articulation of the *coniunctio oppositorum,* the soul's paradoxical healing state in the union of opposites. This goes deeper than subject and object and is an experience of *anima mundi.* In the symbolism of alchemy, he showed that treasures such as gold or precious gems become images not only of wealth but of the soul's most valued state, wholeness. Archetypal theory is in the background when we speak of technology as having mythic or religious significance beyond ego and opening a path to *anima mundi.*

The German archetypal psychoanalyst Wolfgang Giegerich sees technology not as a rational undertaking but as a style of unconsciousness:

> "Technology is not the opposite of instinct and unconscious psyche but another style of unconscious. It is not the case that our consciousness has become rationalistic, calculative and so on; in point of fact, it is still very much entangled in idealistic, nostalgic, sentimental attitudes. What has happened is that instinctual psychic life has changed its language and its medium from a mythical to a rationalistic style. The objective [collective] psyche today is represented by technology. Technology is our nature, our new earth, our instinct, our body, our spiritual, symbolic life." (Qtd. in Avens)

Robert Romanyshyn also wrote a valuable archetypal study, *Technology as Symptom and Dream,* where he explores the self as spectator, the abandoned body, and says: "[T]he robot is an image of the industrial worker" (145).

Jung explored the *unus mundus,* the one world, "a potential world, the eternal Ground of all empirical being, just as the Self is the ground and origin of the individual personality past, present, and future" (Jung, "Conjunction," para. 760). The theoretical instants before the Big Bang are the *unus mundus,* when everything was a united whole, yet the seeds of the universe's multiplicity had to be there, in correct detail. This is a transcendental mystery, so we cannot claim too much certain knowledge of it. Discussion of the union of opposites cannot take dogmatic or literal form. It must restrain ego's yearning for certainty and remain in the realm of the poetic, symbolic, and mythic language of the *unus mundus.* This is why so many raw religious experiences are ineffable, poetic, and paradoxical. Understanding them requires a return to a primal state of transcendental consciousness underlying this world's opposites. In various cultures this is called the realm of the Kingdom of God, the Tao, Enlightenment, Emptiness, the Force in *Star Wars,* or Being.

Technology and Environment

Technology is obviously shaped by the environment, which supplies the concrete, wood, glass, steel, oil, and silicon for making machinery. Industrial societies conveniently see in the environment an unlimited reserve of raw materials for human technology, but the late twentieth-century ecological crisis inaugurated a new phase of recognizing the limits of that kind of exploitative consciousness. This problem was not even dimly imagined by the founders of industrial culture, whose new machines had little impact on the environment. It wasn't until the smoky "satanic mills" began filling urban air and water with filthy toxins that the scale of technology's impact on the environment became evident. Today, the area around the Chernobyl and Hanford, Washington, nuclear plants are two of the most radioactive spots on earth. These are hardly neutral spaces inhabited by life-enhancing technology; they are parts of earth's life-supporting systems, too long neglected, now seriously damaged. Overfishing our oceans and polluting air, water, and soil have also been the results of technological success, and restraint has been strongly resisted because so many corporations and individuals are enamored with the power and riches of industrialism's wizardry, which requires ignoring the environment. This is a major, dangerous enchantment of technology, with deep psychological roots that urgently need re-visionings.

Jean Brun's Technology of Desire

Jean Brun, formerly a professor of philosophy at the University of Dijon and author of *Les Masques du Désir* (The masks of desire), was a forerunner of enchantment theory. He argues that:

1. Technology is a human effort at self-sufficiency that must reject the theology of a transcendent Creator and the presence of Being-in-the-world.
2. The development of subjectivity as the focus of knowledge and authority in the technological world has become inflated into a way of ego's grasping for absolutes that in effect seeks the powers of divinity for the subject.
3. The classic subject/object dichotomy, fundamental to industrial society's metaphysics, is artificially constructed and not self-evident or comprehensive.

Brun's thought is influenced by Gaston Bachelard, an influential French phenomenologist at the Sorbonne. In *Les Masques du Désir,* Brun merges the insights of the psychology of the unconscious with the philosophy of phenomenology. When he opens the cellar door of the unconscious soul, he calls the contents "desire," a broader term than Freud's libidinal "id" and sometimes as archetypal as Jung's "collective unconscious."

Brun perceives the enchantment of the world just under the surface of the technologies that disenchantment built. The archaic immersion in nature, with its mystical visions, symbolic icons, and shamanic trances, has not faded away nearly as much as Gauchet and others envision. While disenchanted consciousness works away on the stage of technological culture, dazzling us with new inventions, just behind the scenery we can find unconscious desire quietly pulling the ropes. Technology is animated by desire, Brun argues, and the masks concealing desire must be torn away. Too many technologies, presented as controllable tools of liberation, have become weapons of death. Myth has not been dissolved by reason. Myths are the most powerful moving forces behind technology.

Brun explores the history of the use of fire, flight, magnetism, electricity, gunpowder, and robots and finds mythic dramas of humanity seeking to embody desires. The rational subject that sees itself in control of the machines does not acknowledge its own passions. Myths are not errors and untruths, as in the disenchanted worldview, but symbolic, meaningful stories, powerful motivators in the industrial culture that minimizes their significance. Brun draws primarily on Greek myths, pointing out the footprints of Daedalus, Prometheus, or Dionysus: "Prometheus no longer contents himself to work on Nature; he now searches to work on himself and to become his own tool and his own material" (*Masques* 187).

Humanity cannot be defined exclusively as a tool-making animal, Brun argues, for desire is ontologically deeper, guiding the process of tool making: "The first *mask* of this unconsciously redemptive desire is that of the *homo faber,* which is repeatedly promoted as the father of *homo sapiens*" (9). Like Marcel Gauchet and Jacques Derrida, Brun relativizes the subject/object framework of technological culture:

> The "subject" is "subject" only because it is relative to the "object." The "object" is "object" because it is relative to the "subject." . . . All dichotomy finds itself thus rejected, the dichotomy of subject-object as well as the dichotomy of exterior-interior, or the dichotomy of appearance-reality. (203)

Brun sees technology as striving to gain mastery by pulling humanity away from its immersion in a primordial field of desire. Desire is not satisfied with temporal power but, like Prometheus, seeks transcendence: "[T]he Technology of Desire constitutes the permanent temptation of humanity to become its own God" (228). But the atheist philosophy of technology cannot admit that it seeks divine powers, so it masks this desire (8).

Not only is technology a quest for self-sufficiency masked as divine absolute power, but, as a product of desire, it knows no limits: "Technique is not self-regulating, because Desire, of which it is the instrument, animates humanity entirely" (9). Progress vaguely knows its roots but fails to interrogate its own goals. Progress is thus used to justify constantly expanding "economic" growth with no restraint and thinly veiled goals such as self-aggrandizing pleasures and damaging powers of the subject exploiting objects.

Desire is a dynamic, passionate master, seeking not only divine absolutes and dominion but ecstasy, echoing Dionysus (8). Yet this drama of desire's technologies creates a world whose very style of "objective" consciousness causes its own suffering:

> A person is not neurotic because he has such and such a disease which plunges him into his unconscious conflicts; humans are neurotic because our entire consciousness is a disease. (15)

Uniquely modern anxieties, stresses, and neuroses are not aberrations of the dreams of technological progress; they are the suffering caused by its greatest desires.

Logos/Mythos

Plato clarified an old continuum between *logos*—logic and reason, which he developed brilliantly—and *mythos,* the imaginative stories of the gods. This distinction hardened into a dualism in western thought. Christians insisted that their faith is a historically based *logos,* while other religions are a false *mythos.* But the flaws of such a dichotomy were avoided to some extent in Plato's dialogues. He frequently develops a logical train of thought that reaches a dilemma, then he turns to myth to resolve it (see, for example, *Gorgias,* sec. 523). In addition to his careful reasoning, he develops an elaborate mythic cosmology of eternal souls (*Phaedo,* 71a–115a). In *The Myths of Plato,* J. A. Stewart argues that myths for Plato represent a priori assumptions, not disposable legends. Aristotle developed *logos* further but also put his whole system under the theology of the divine unmoved mover (*Metaphysics,* book 7, chap. 7). Most Christians still deny the mythic roots of their faith, such as the goddess elements of Mary's virgin birth. Today, techno-society still neglects the *mythos* hidden in the *logos,* such as God's blessing, in Genesis, on the human domination of nature.

To emphasize the role of unconscious enchantments on conscious reason-

ing is not to deny the important and powerful role of reason. Plenty of reasoning is used in this book. Technologies need the precise logic that modern
reasoning has developed. However, there is a huge gap between the technical
use of reason and the exaggerated claims that reason alone is sufficient to know
everything, because it is not in itself completely logical. It relies on a will to
power that is an enchantment just below the surface, not a rational argument
itself. Similarly, there is an important difference between science as an inves-
tigative method and "scientism," the utopian, triumphal claim that science is
sufficient to know everything and that other methods, such as intuition, are
invalid or inferior.

A positive, cooperative working relationship between reason and intuition,
imagination, and poetic language is necessary to incorporate the wide spectrum of human knowledge. More human faculties are needed to make our
experiences conscious. Reason may trick itself into imperious claims (such
as the inevitability of techno-progress), just as enchantments can (such as
the rightness of patriarchy). Each may criticize the other or work together
to reach agreements on truth. Thus science (*logos*) should not be seen as the
automatic standard for all truth, nor should arts (*mythos*) be seen as mere
"entertainment."

Desire's Myths of Technology: Motors and Gunpowder

The study of mythology has burst into modernity like a surprise guest at a party.
Since archetypal psychologists have showed the meaningfulness of myths as
archetypal stories emerging from the collective unconscious, Joseph Campbell
and a host of other scholars and artists have brought myth back to consciousness as a symbolic legend full of enchantment. As Thomas Moore says,

> Mythology is a special tool for enchantment, because it brings to mind the
> world of invisibles—the spirits, thoughts, and emotions that crowd our imagi-
> nations . . . giving full articulation to matters that can't be measured—things
> like love, hate, death, fear, and evil. (*Re-enchantment* 233–34)

Greek gods such as Daedalus, Icarus, Prometheus, and Hephaestus are "obviously models for *homo faber* because Faustian man from the beginning
gave birth to them" (Brun, *Masques,* 19). Brun's view of myths draws on
Jung's theory of archetypal images symbolized in myth becoming ideas and
technologies:

> The dreams of humanity represent archetypes of clear consciousness as much
> as the archetypes of a nocturnal unconscious. They are the embryos from

which are born elaborated systems . . . the powerful motor of epistemology. . . .
[M]yths are messages and interpretations, sons of human secrets; they become
fathers of ideas and actions which transform these secrets into programs of
scientific research and historical conquests. (20)

Myths are symbolic expressions of meaningful existential concerns, and they are
naturally transformed into technologies that feed on and reproduce them:

> Behind the powers that technology puts at our disposal can be found all sorts
> of existential concerns. . . . These concerns are as much attempts to pierce
> the mystery of Being, and consequently they are about birth, life, love, and
> death. It should not surprise us, then, that by the end of the twentieth century,
> technology and science had finished by becoming, with history, very powerful
> at generating myths. (21)

From H. G. Wells's *Time Machine* to *Star Trek, Star Wars,* and *The Matrix,*
science fiction is the premiere mythic genre of our technological culture. It
is usually interpreted as if it were a prediction of future technical prowess,
rather than mythology, but it is the mythic language of technological desire
and its current fantasies:

> [Science fiction] presents itself to us as the grand liberator from myths, as the
> expression of *Logos* having succeeded over *Mythos.* [But it] has become, with
> history, the most powerful mythogenetic factor of our time. (220)

Of course, this literature is highly ambivalent, showing the victorious hero on
the one hand, bringing salvation thanks to bravado and lots of machinery—fast
spaceships and powerful weapons—while on the other hand picturing the
terrors of technology gone wrong in a fiery mushroom cloud, vicious aliens,
or robotic tyrants. The devastating irony of dreams of liberation bringing
nightmares of devastation is rooted in the denial of the influence of desire in
technoculture. Technology's desire-filled mythic heavens are haunted by its
hells, with only the rickety promise of faith in progress to pull it out of despair.
Horrendous aliens, vicious tyrants, totalitarian robots, and nuclear fireballs
express modernity's painful nihilism. This deep moral uncertainty about the
outcome of high-tech culture is the most painful of technology's enchantments, often expressed in explosive violence. Acknowledging this shallow
moral ambivalence is a step toward healing this enchantment.

Brun pulls the deeper patterns of mythic desire from specific technologies.
Behind the mechanical wheel he spots the sacred mythic chariot, which in
Greek and Hindu mythology was a cosmic model, an ontological picture of
the heavens and earth revolving around a central axis (30). The incarnation
of the cosmic wheel into a machine was a heroic act: "The first maker of a

wheel was therefore a man who accomplished a sort of Promethean prowess, a sacrilege for certain, grandiose for the most part: he brought the wheel from Heaven to Earth; he installed the supernatural in nature" (31).

Gunpowder and fireworks can be seen as a "domestication of celestial fire" (46). Early astonished reactions to gunpowder explosions were to the tremendous noise and light, which sounded like thunder and lightning. Roger Bacon said, "'One can imitate thunder and lightning in a manner more terrifying than in nature'" (qtd. in Brun, *Masques*, 47). The fantasy that these celestial powers had come under the control of one's military opponent at first made them seem divine. European rifles were reportedly called "thunder tubes" by some Native Americans. According to Brun, "Thanks to the invention of gunpowder, humans grabbed lightning from the hands of Jupiter" (47). How many soldiers using gunpowder against opponents with spears resisted the desire to feel absolutely powerful? What primordial and cosmological desires are released in holiday fireworks, cinematic explosions, and the theory of the original Big Bang?

Desire Masking as Technology

Brun shows that the world is not as disenchanted, or stripped of desires, as conventional consciousness assumes and that desire fills the supposedly mechanical world with compelling mystiques. One barrier to open discussion of this idea is the denial of concealed motives. Desires such as pride or fear mask themselves as machinery. Arguments about technology always bear heavy loads of unstated value judgments. They are frontmen for concealed, unacknowledged motives, Arnold Pacey says: "The overriding 'motivation' in nuclear development has nearly always been 'pride,' especially *pride* of the sort called *patriotism*" (*Culture*, 90). Pride (whether proper self-respect or conceited arrogance) and patriotism (whether appreciation of one's country or fierce, intolerant aggression) are examples of desires in conventional discourse. Too often they rear their pious, ferocious heads and not only interfere with, but stubbornly guide rational analysis. Again, this shows that technologies are not simply directed by conscious ego willfulness. Machinery becomes a powerful logical structure with passionate imperatives, embodying strong desires, but the polite neglect and suppression of concealed motives prevents awareness of deeper motives that drive the discourse. One of these is gender.

Women and Technology

Feminists argue that, in the history of technology, the patriarchal subordination of qualities defined as feminine, whether they are actually restricted to women or not, has been a strong dynamic. Carolyn Merchant investigates this theme in *The Death of Nature: Women, Ecology, and the Scientific Revolution*. She shows how mining was for a long time seen as a violation of Mother Earth. As Edmund Spenser wrote in *The Faerie Queen* (1595), about the miner,

> "Then gan a cursed hand the quiet wombe
> Of his great Grandmother with steele to wound,
> And the hid treasures in her sacred tombe,
> With Sacriledge to dig. Therein he found
> Fountaines of gold and silver to abound,
> Of which the matter of his huge desire
> And pompous pride eftsoones [again] he did compound." (Qtd. in Merchant 38–39)

Whereas ancient images of Mother Earth pictured her actively promoting life, as in the Babylonian myth of Ishtar's descent, by the Renaissance she is seen more as passive and docile, surviving increasingly violent abuse, rape, and greed. Doubtless the gold and silver mined so rapaciously in the New World increased this shift in consciousness. Images of avarice, impiety, and lust filled the literature that was critical of mining. John Donne penned an erotic vision of New World mining in his Eligie XX, "To His Mistress Going to Bed":

> "License my roaving hands, and let them go,
> Before, behind, between, above, below.
> O my America! My New-found-land,
> My Kingdome, safeliest when with one man mann'd,
> My Myne of precious stones, my emperie,
> How am I blest in this discovering thee!"

Hardly the picture of a neutral, objective technology, mining was rife with fantasies of invasion of female privacy and rape. This masculine view of nature pervades the notion of objectivity, a problem that remains in the collective unconscious of science and technology, especially when it is denied.

Similarly, women's control of their own reproductive realms was ripped away from them. Instead of inviting midwives to study scientific medicine and add medical knowledge to their traditions, men fiercely persecuted them. "Forceps" were invented by men to aid in births, while midwives were burned as witches. The struggle by men to gain control over women's bodies was part of the overall effort to gain control over nature (Merchant 155). Again, masculine invasion of femininity was a forceful, patriarchal ideology, not an objective trend.

Donna Haraway adds force to this argument by showing how the early images of science as a manly undertaking like war aggressively excluded women and their views from technoscientific developments ("Modest" 27–35). The early British Royal Society excluded women. Race, gender, and class barriers were strictly enforced for a long time in deciding what counted as valid knowledge. In the effort to develop an epistemology of "objectivity," women were relegated to the realm of "subjectivity" and still suffer male distortions and suppression in technoscience. "Nature" itself is a projection full of passions, from rape to war.

Men even quoted the Bible to prevent women from inventing and patenting devices. Despite such widespread suppression, a few creative women began patenting devices after the U.S. patent law of 1790, which was silent on gender (MacDonald 4, 10). As Anne MacDonald describes in *Feminine Ingenuity: How Women Inventors Changed America,* most early women's inventions were domestic and clothing improvements, from hats to brassieres (Crosby, *Passionate,* 72–74). Some nineteenth-century women invented non-domestic devices. Martha Coston refined pyrotechnic night signal flares in 1871 (MacDonald, 16–19). Mary Meyers was an aeronautics pioneer who owned a popular "Balloon Farm" in New York and experimented with many flying devices, such as balloon steering and parachuting in the 1880s (157–59). In dry Nevada, Harriet Strong invented a water-conservation technique using dams (159–66). In 1965, Dr. Stephanie Kwolek, a DuPont chemist, invented the chemical in Kevlar, used in bulletproof vests (373–75). By the twentieth century, the growing list of women's inventions shows that the male suppression of women's ingenuity with technology was an unjust fantasy.

The Compelling Mystique

Obviously, invention is far from an objective undertaking, since it involves determination and creativity as well as technical skill. Inventors often attribute their inspiration to unconscious sources. The organic chemist Friedreich Kekulé had a dreamy vision of a snake eating its tail that inspired his correct hypothesis of the benzene ring molecule (Russell, *Valency,* 70, 242–43). The common refusal to acknowledge suppressed imagination and desire allows a technological metaphysic to build an accepted worldview with an autonomous life that conceals its mystique. But technology is not simply a rational tool that we use for conscious goals but it soaks the entire techno-worldview wet with passions, pleasant and painful. An enchanted framework overwhelms theoretically rational intentions. As Erik Davis says,

> [T]echnology installs a new and invisible framework around the world we live in, a potentially catastrophic structure of knowing and being that swallows us up whether we like it or not. . . . The immense machineries of war or entertainment can hardly be said to proceed from rational necessity. (144, 10)

"Magic" in a scientific world is interpreted as a peripheral, harmless entertainment enacted by stage performers doing explainable tricks. But even materialistic science and technology are cloaked with the mantle of magic. Electricity, speedy vehicles, dramatic communications technologies, and supersonic flight embody many of the dreams of magic (Davis 38). But Freud, Brun, and others point to an ontology of desire that threatens the monolithic materialism of the West. Science and technology have not despiritualized the world nearly as much as we think, for they have enacted dreams such as flight, once called magic while suppressing natural wonders.

When we look at the large gaps in our knowledge, unexplained aberrations, and rapid changes in quantum physics, we need not readily accept the claim of the universal explanatory power of science. Most people still imagine the world in a mechanical, materialistic way that has been outmoded by new relativity and quantum theories in physics. Large gaps remain in the lag between those theories and daily life. Invisible "dark matter," for example, seems to fill 90 percent of the universe, astronomers say, but we know very little about it. Two of their proposed names for the mysterious stuff are enchantingly titled with masculine-fantasy nicknames: "Machos" or "Wimps" (Hazen; Chandra). So how can anyone claim certain, complete knowledge? The dance of energy that holds the world together is an overwhelming mystery. We know only a useful fragment.

The fascinations and mythic visions of technology in science fiction, for example, are a like a self-imposed mass-hypnotic fantasy. The power and grandeur of our enchantments with technology are a compelling mystique, not a rational framework. In *Myths That Rule America,* Herbert London and Albert Weeks see in technology a hypnotic extension of ourselves that lets us down. The myth of technological rule constantly regenerates the illusion that anything is possible. Technology seems to offer ultimate salvation from all suffering, but instead it sucks us into an addictive, druglike dependence. We blame our addiction on the machines instead of taking responsibility for placing our fate in mechanical hands. Increasingly dehumanized, we let machines become our collective will, expecting them to solve problems technically that instead require social responsibility. The mystique of machinery ensnares our imagination, hypnotizing the collective will into the grand fantasy that machines can solve all our problems. One letdown after another is ignored while the illusions keep on breeding new dreamy expectations (London and Weeks 33–34).

The enchantments of technology are a kind of mysticism. Of course, this is a heretical notion to modern technologists, but heresy is no stranger to the search for truth. This location of mysticism at the heart of technology is simply a recognition of the depth of mystery in technological beliefs and enchantments that has been denied and suppressed for too long. In *God and the Chip: Religion and the Culture of Technology,* William Stahl sees a technological mysticism holding together the core of technological culture:

> As practice, identity, and mystification, technological mysticism lies at the heart of advanced industrial society. . . . Like a religion, technological mysticism 'binds together' core values into a coherent, if implicit (and often unexamined) set of beliefs and rituals. But do we want to accept it as the One True Faith? (19)

"Magical," "mythic," "hypnotic," "mystical"—these are hardly expected descriptions of technology, but these words express a new layer of understanding technology as part of human culture, with all its passions, beneath the sophisticated rational analysis. The imaginative and hypnotic dimension of thought is so fundamental that there is no question of eliminating it as such. It has always been there. The question is what *quality* of depth and mystery will prevail and command our faith? Religions have always expressed a culture's most absolute (and hopefully high-quality) beliefs, values, social frameworks, and ontologies. If consciousness is no longer considered to be ruled by cognitive reason but to have imaginative, mythic, dreamlike, and even spiritual enchantments operating the controls far more than it thinks, then its rationality and the certainty of its metaphysical assumptions are thrown into doubt.

When we question the unthinkable—the fundamental metaphysics of the subject/object dichotomy—the disenchantment of the world is over. The enchantments that were always concealed behind this dualism are waiting to be released from their repression. Consistent with the finest benefits of individual, political, and technological freedoms and rational thought, new paths are needed for imagining technology. Many of the elements of traditional religions are irrelevant or hostile to this task. But a renewed spirituality can incorporate the best of technological practice, such as an earth spirituality behind deep ecology that places humans within nature as participants, not as outside observers and exploiters. The effort to determine the finest elements of technological culture is an ongoing cultural debate. In the chapters that follow, we will examine some important enchantments, such as the subject/object dichotomy, triumphalism, speed, utopias, space travel, and robots.

2 The Obtuse Object

> The concept of a neutral world, untouched by man's efforts, indifferent
> to his activities, obdurate to his wish and supplication, is one of the
> great triumphs of man's imagination, and in itself it represents a fresh
> human value.
>
> —Lewis Mumford, *Technics and Civilization*

Binary Thinking

Not infrequently in the history of thought, some concepts come to seem
more real that the concrete realities that give rise to them. Such is the case
today with the notions of object and subject. Human bodies are binary in
structure, with pairs of arms, legs, eyes, ears, and a somewhat binary brain.
Bodily structure must influence our thought—left/right, hot/cold, male/female.
We have ten fingers, with which children learn to count. Is it by chance that
we have a *digital* ten-based number system? Why have we divided the world
into two fundamental parts and called them "subjects" and "objects?" Why
do we have a true/false epistemology and a right/wrong morality when we
constantly find problems in the large intermediate areas? Even the life/death
pair surrounds a large grey area called near-death experiences. Yet we cling
to these simple dichotomies that make thinking easier. But is reality actually
structured in binary forms? That would be convenient, but is it true? We must
honestly explore other options, such as paradox and continuum, with the risk
of deconstructing comfortable formulas.

Neutral Thingness

The concept of the object stands for the interlocking set of premises that define
our relations to nature as *obtuse*—blunt, opaque, and detached. To imagine
a pond, a tree, a mountain, a person, a machine, or a work of art as an object
is to imagine oneself standing outside it and to deny it the presence of more

than obtuse matter. The naturalist author Barry Lopez says, "If you behave as though there were no spiritual dimension to the place, then you can treat the place like an object" (60). This stance fragments the world by selecting its mechanical aspects for technological use and ignoring or denying its soul, its inherent value, and its spiritual implications.

In industrial societies, surgeons work on a patient's body as if it were a neutral thing that we call an object. They suppress their emotions to operate effectively as healers without fear, revulsion, or grief. This is beneficial, but there are many negative consequences of such suppression. Pilots of military bombers, for example, detach themselves from the human and natural world below in order to perform their missions. Industrial consciousness, whether capitalist or socialist, treats the world as a set of dead objects, or resources to be exploited for profit, with little regard for the side-effects. Huge fishing ships sweep vast numbers of oceanic life forms into massive nets, discarding a large percentage of dead ones as commercially undesirable, as if there were an unlimited supply of such objects with no environmental significance. Land is treated as "real estate," a neutral object with no significance beyond market factors and legal protections, as if parts of the earth could be owned by passing inhabitants. This mechanical world picture imagines things as if they were neutral, soulless external objects severed from observing subjects.

What is an object? Literally, the word denotes something "thrown before." The etymology is rooted in the Latin *obiecto*, "to throw in the way, or reproach," as in a court objection in opposition: "I object!" *Obiecto* originally meant not a material thing standing independent of a subject but rather an act of blocking. Also, the linguistic usage as the object, or predicate, of a verb led it to mean a quality or a goal to which an action is directed.

Today the meaning is the opposite of the original. Since the Enlightenment, an "object" has generally come to be understood in two major ways:

1. First, it is understood as a *material thing* perceptible to the senses, ruled by the laws of physics, standing opposite a "subject." An object is theoretically observed and evaluated only by rational, quantitative criteria. An object in this metaphysic is part of the Cartesian project of stripping the world down to lifeless, inert thingness with no inherent value and no meaning beyond what humans perceive in it and give to it.

Thomas Hobbes gave a classic definition: "[The thoughts of men] are every one of them a *representation* or *appearance* of some quality, or other accident of a body without us, which is commonly called an Object" (85). This notion of thoughts as simply representations of objects has since lost its force, but the idea of the object as a "body without us" remains. The purposes that industrial society now gives objects are part of a plan in which they can be dominated and exploited for maximum technological and commercial power. The "object" was originally conceived as the result of stripping the world of its enchantments with ancient magical and divine powers. But objects have not become neutral things. They are still enshrouded with human enchantments: the modern drives for technological power and pleasure.

2. Second, the "object" is the standard for truth in the industrial world. Humans must conform by thinking "objectively," that is, by keeping "subjective" feelings such as pride, competition, and irrelevant biases out of science, being fair to opponents, not letting political views interfere with reasoning, and considering different, even opposing, perspectives in a debate. When objects are misinterpreted by excessively subjective feelings (as in the crude concept of "sex objects"), psychology calls that "projection." In psychological theory this subjectivity, which is projected onto objects, must be "withdrawn" before a purely rational, objective analysis can proceed.

This method of determining truth developed in opposition to symbolic superstitions (such as sacred trees), pseudoscience (astrology), religious dogmas that sanctioned outdated metaphysics (the earth as the center of the universe), and limits on scientific exploration (no anatomical dissection). The scientific method of knowledge undertakes empirical tests of objects to determine the validity of hypotheses by rational analysis, preferring quantitative, objective data.

Many aspects of the "objectness" metaphysic are obviously valuable and indispensable, especially in the realm of technology. But objectivity has been promoted as self-evident and the universal exclusive truth in industrial culture, even though it has some seriously negative consequences, such as fantasies of neutrality and control. The limitations on technological control (as in nuclear radiation) show that we are too confident in our enchantment with domination of external objects, which do not just sit there, waiting to be controlled. The time is ripe to question faith in the metaphysical model of the supposedly neutral object and the mental state of objectivity.

"Objectivity" is far from objective. The standard of a completely neutral thingness existing independent of human understanding, by a neutral, feelingless attitude of establishing facts, is unreachable. Medical research, for example, is funded to a great extent by pharmaceutical and medical equipment industries, which threatens the objective value of the research. What we call "objectivity" is full of nonobjective, irrational passions, desires, and purposes that are taken for granted. A fact is a part of the framework of objectivity that seems to establish certain truth, and there are obviously times when facts are needed. But the fact is not what it is dressed up to be.

"Fact" as a Fragment of Truth

The word "fact" has been asked to carry the weight of the authority of the world of objects. In theory, a fact is information known with certainty, objectively verified, having real, demonstrable existence. But as any journalist or politician knows, facts can be stated in ways that distort or ignore the larger context that gives them meaning. Facts can be half-truths or truths that ignore other essential, embarrassing truths. Facts always need a context to give them meaning. A fact is merely a fragment of truth. And the context of facts is not only more self-evident and autonomous bits of information but a worldview, even a dramatic mythic vision, such as the belief in humanity's role as dominator of nature.

Factuality is obviously valuable when we seek a fair, unbiased understanding. But ultimately that is not possible, because human purposes and goals are always involved in any statement of fact. People seek survival, health, control, safety, political dominion, and personal achievement. In this worldview, some agree to neglect major environmental considerations and issues of social justice and to ignore or deny death, ethical issues, and spirituality. These are all theoretical positions in which many so-called facts are embedded, for better or worse. This is not a counsel for radical relativism, nihilism, or "anything goes." Standards of truth and rightness are obviously needed, but they cannot be determined solely by the objectivity model. This frightens or angers some people who are fully invested in this model. But these fearful, angry feelings are evidence of the limits of the goal of objectivity, which in theory would be unconnected to feelings. But obviously we are all human, and the problem is the *quality* of feelings and theories involved, not only quantitative or logical information. What we think of as objectivity is actually a spectrum of mental phenomena ranging from detached neutral observation of nature, through a collective social consensus called "truth," to a concealed aggressiveness rooted in the human passions.

Under the surface of industrial consciousness, the very nature of what we label as "objects" is greatly determined by the thinly veiled charm of the technological enchantment, especially the blunt drive to increase human power and dominion. Objects are conceptually the obtuse philosophical slaves of the material world, selected slices of reality forced to serve human dominion. They are obtuse because we see them darkly, only through selected conceptions, ignoring other factors such as the environment that makes objects possible. Objects, seen only as thingness by the forced externalization of "irrelevant" considerations, then produce "unexpected consequences" such as pesky nuclear radiation. What deeper elements lie at the roots of this construct?

Hunter-Warrior Detachment

Usually the mental skills we call "objectivity" are portrayed as arising from the faculty of reason developed mainly by the Greeks and Enlightenment thinkers. Objectivity is typically distinguished from instinctive behavior and maintains an attitude of detached, neutral rationality against instincts that might bias and distort the objective quest for truth. But I suggest that the roots and dynamics of what we call detached objectivity are actually more complex and indeed instinctive in themselves. Perhaps objectivity, the heart of scientific and technological consciousness, is the refinement of an unconscious archetypal pattern, namely the hunting instinct. Our technological culture is saturated with the submerged instinctive depth of the hunter, which has evolved into the refined skills of objective reason on the surface. Psychologically, objectivity has deep roots in the ancient detachment of the hunter from the horror of death.

Any archetypal mental impulse will develop both positive and negative aspects, like the hunter who provides sustaining food but must also deal with aggression and death. The hunter is courageous, skillful with weapons, and willing to encounter the defenses of dangerous animals. The hunter must also be strong enough to butcher the bloody carcass without discomfort and take responsibility for killing another living creature. This instinct is deeply rooted in our predatory animal drives; archaic hunters were frequently identified with predatory animals such as lions or panthers, and thus ancient art from Greece to Central America depicts heroes and kings wearing animal skins or heads.

Archaic people were not always crudely aggressive toward hunted animals. Religious studies have shown that the attitude of respect and reverence for the spirit world that provides animals for food was blended with killing skills. Archaic hunters frequently had to undergo rigorous purification rituals before the hunt, to make themselves worthy of taking a life. After killing, they may have offered some of the blood to the species' ancestral power that provided the animals, to ask that the species remain vital and a new animal replace the killed one. Hunters often dressed as animals and imitated them in dances or in a totemic identification rite in order to "become" the animal in the hunt.

But, as we see among the Australian Aborigines, a hunter was typically not permitted to kill his own totemic animal, since humans and animals are related. These rituals provided various ways of transforming the horror and guilt of killing another being into a mythological relationship that sought to maintain life's renewing powers. Even so, the aggressive, detached approach to hunting must have also been a factor in building up the courage to endure the difficulties of the hunt. Imagine the mental state required to shoot a poison dart at a monkey or to stampede buffalo over a cliff. Later, as technology improved, iron weapons increased aggressive conquests and military might. When gunpowder was introduced, detached killing at a distance became much easier. Hunted animals could become more like objects.

The hunter of animals became the warrior, hunting the enemy. The detachment from compassion for animals and the enemy was eventually elevated into detached observation of the world as a whole. Objectivity was not a new mental skill that was cleverly invented by modern rationality; the objective attitude did not sprout newborn from the brains of Enlightenment philosophers. It grew slowly out of the dark shadows of the ancient archetypal instincts of the hunter-warrior. Its modern form has beneficial aspects, such as fairness and neutrality, and dark aspects, such as cold, sometimes cruel detachment.

The modern hunter is boldly aggressive toward nature, seeking out prey and suppressing sympathy to succeed in the kill. The scientist who must suppress sympathy for life in order to dissect an animal in a biology lab must reach deeply into the collective psyche to stifle the same compassion for the refined purpose of learning anatomy objectively and thus understanding life more fully. Like a hunter, the technically skilled military pilot, rifle shooter, or submarine torpedo targeter must learn to hate, then objectively repress sympathy for the victim while aiming to kill. The hunter's repression of sympathetic feelings is required of the scientist or engineer who sees the world as comprised of neutral objects that are more easily dominated than those with emotional or ethical attachments.

Of course, invocation of the concept of objectivity has led to several admirable aspects of modern industrial culture. It has promoted the refined ethic of fairness, neutral judgment, and collective verification in laboratories, schools, and courtrooms. The goal of neutrality has greatly aided in reducing bigotry by bringing people of various ethnic and cultural backgrounds face to face. But objectivity has not succeeded in suppressing its shadow. The hunter-warrior instinct is a union of positive and negative drives that easily mingle and struggle for dominion. Appeals to objectivity as an ethical arbiter for industrial society's ills may suffer from concealed hostility, as in the way the Superfund for cleaning up toxic waste in the United States has been diminished by legal battles over who is responsible for causing the problems. More money seems to have gone into legal battles and public relations than into actually cleaning up the toxic dumps. The fighting spirit stampedes the objectivity of a neutral assessment.

There is no such thing as an object or objectivity. What we call an "object" is not a material thing in itself but a theoretical construct of a materialist or realist metaphysic that tries to strip the world down to a lifeless structure with no inherent value. Nor is objectivity the standard for detached, neutral truth that it proposes to be, because it cannot purge itself of passions such as "sex objects" that mingle with its theoretically value-free rationality. While the use of the "objectivity" metaphysic and epistemology in the scientific method has produced a powerful technology and some valuable methods for seeking neutral thought, its method is far from complete, so it cannot explain everything. Most embarrassing, this method of thought must resort to a deceptive denial of the unconscious desires that it cannot expel from its own house. The very detachment at the heart of the method of objectivity is its downfall, because it promotes this denial of the fullness of experience. The unexpected consequences emerge as ignored but costly side-effects, such as immensely expensive cleanups of dangerous toxic waste.

Punchkin's Parrot

A deeper layer submerged below the modern attitude of objectivity is expressed in a widespread folktale motif regarding the External Soul. The theme is that a person has deposited his or her soul outside the body in a safe place, to protect his or her life. The person may be an ordinary woman, a beastly tyrant, a folktale ogre, a heroic warrior, or a newborn child. The soul may be deposited in a secret place far away, protected from discovery, or it may be placed in a commonly known animal or tree. The soul may be located in hair, an egg, light, a vital organ, a gemstone, crocodile, shark, fig tree, coconut, or bird. The external soul is conceived as a brother or sister, an alter ego or double (*Doppelgänger*), and as long as one of the pair lives, the other lives. But if one dies, the other also promptly dies. Thus it is believed that if one could protect the external soul from death, one could gain the power of immortality (Frazer, *Golden,* chaps. 66–67, 773–802).

In an old Hindu folktale this power is abused by the magician Punchkin, who is determined to marry a queen and holds her captive for twelve years. She constantly rejects him, and finally, when her son comes to rescue her, they plot Punchkin's overthrow. The queen pretends to consent to marriage and asks him coyly, "Are you quite immortal? Can death never touch you?" He proudly replies, "It is true, that I am not as others." He describes how, thousands of miles away, in the midst of a thick jungle, is hidden a cage containing a parrot. "If that parrot dies," he confides, "I will die. But no one can kill that bird, for thousands of genii surround the remote place, guarding it for me." The resourceful prince captures the parrot and returns to the door of the magician's palace with it. Enraged, Punchkin demands the bird, but instead the boy rips off a wing. As he does so, one of the magician's arms falls off. The torture continues until the parrot's neck is wrung. At the same instant, the cruel magician's head twists around, and he expires.

When the goal of the placement of the soul in the external environment is to avoid suffering, to overcome death, and to gain immortality, the very placement of the soul "out there" in a detached place grants one a strong ego but also turns the person into a victim. Externalized souls suffer from this detachment, not only because the soul is remote from the person who externalized it, but also because it can be captured for destructive uses. The vulnerability of those who hide their soul is evident in their need for numerous defenses, like the thousands of genii guarding the parrot's hiding place.

The magic, or enchantment, of detaching and placing the soul out into the world requires numerous defenses. Like in the story of Punchkin, the fragment of soul remaining becomes an inflated ego, claiming to work magic but needing to exert force to attract love. It fails at love because it becomes arrogant instead of intuitive, sensitive, and capable of the full range of feeling. Punchkin must hide the sensitive soul deep in a wild (unconscious) jungle and guard it defiantly behind defenses. His strength may allow him to create powerful external technologies—cars, rockets, or robots—but, like a cruel tyrant, the magician is insensitive and blundering when it comes to deeply desired goals such as love.

The externalization of the soul creates not a neutral, completely truthful realm of technological objects but an unconscious jungle concealing a precious soul guarded by many defenses. And the detached magician becomes a soulless, power-hungry tyrant who is numb to love. This folktale tells the story of too many technological wizards today who think that they are working with objects out in the world. But, by narrowing their focus to technical thinking, they hide their sensitive, valuable soul deep in a place that needs many defenses. They may put the imagined world of objects in the service of defenses and power struggles, but the only way to gain the love that the soul yearns for is to release it from its faraway hidden cage and bring it into daily life. Rather than objective reasoning about detached objects that suppresses bodily and emotional wholeness, the soul yearns to acknowledge the wholeness of the world, where logic and passion intermingle. We can conceptually distinguish them, but they are ultimately united. If we could acknowledge this, we would not suffer from the outcome of technologies that are "technically sweet" but horribly destructive.

Time, Space, and Objects

The religious world always has sacred places of power, whether a star, a mountaintop, an island, or a temple. Sacred times are holy days such as Yom Kippur, Holi, Christmas, or Ramadan, ritually returning participants to the eternal origin of creation to regenerate the present with the power of Being (Eliade 32–33). By contrast, the notions of objects and objectivity are built on the foundational developments of uniform, calculable space and time that culminated in classical mathematics and physics. These abstract techniques built a structure in which the perception of the world as objects free of sacred dimensions developed. Measurement achieved new importance for determining truth, and the human imagination, judgment, and ethical concerns were nearly removed from the world of homogeneous, impartial space and time.

Technologies were part of this development. The mechanical clock refined timekeeping not only into smaller and smaller units but into a different vision of time. As Lewis Mumford has shown, the medieval clock was the monastery or church bell that marked times for prayer as participation in a sacred cosmos. But the mechanical clock removed time into an artificially constructed framework:

> [B]y its essential nature it dissociated time from human events and helped create the belief in an independent world of mathematically measurable sequences: the special world of science. There is relatively little foundation for this belief in common human experience: throughout the year the days are of uneven duration. (Mumford, *Civilization*, 15).

Mechanical town clocks in towers were marks of pride until pocket watches and then wristwatches replaced them. Now digital clocks and atomic clocks provide tiny differentiations of time and are used to finely coordinate electronic instruments and broadcasts. What time is it? A thousand years ago, it was midafternoon. Today, it is 2:35:42 P.M. This mechanical precision does not exactly match the rotation of the earth, which is slowing down, so that in some years a second has to be subtracted from the day of the U.S. atomic clock. The world of motors, computers, and satellites requires a very precise coordination of timing, so the clock is built into the most advanced technology. Nevertheless, mechanical time is a cultural construct, a convenient approximation of nature's rhythms that only seems like an objective, factual system underlying the apparent solidity of the notion of objects.

Space was similarly reshaped from a mythic cosmos into a mechanical cosmos by technique. Ancient painting was largely symbolic and lacking in perspective's illusion of depth. This is evident in archaic petroglyphs depicting visionary experiences or tomb paintings of Egyptian animal-headed gods that used size to depict important figures. Madonna and Child were larger than adjacent humans because they were more important, rather than being large because they were close to the viewer, as in perspectival art. The point was to symbolize spiritual realities, not three-dimensional space.

Renaissance artists developed perspectival art to picture three-dimensional space. Michelangelo's brilliant anatomical studies are a turning point. In his portrayal of God granting life to Adam with the cosmic touch of a finger on the ceiling of the Sistine Chapel, Michelangelo brought together an ancient and a modern vision of reality. He pictured God as the creator of humanity and at the same time portrayed God in a human form, more spatially and anatomically correct than most ancient statues and paintings of gods (outside Greece) had accomplished before.

This technical achievement in art, however, broke the second commandment in Exodus: "You shall not make yourself a graven image, or any likeness of anything that is in heaven above, or that is in the earth beneath" (Exodus 20:4). This commandment is broken repeatedly, but the portrayals of God right in the heart of the Roman Catholic church are an ironic breach that could be tolerated because Michelangelo and Raphael were such brilliant artists, and some Renaissance popes were enchanted by the new vision of three-dimensional space. But to place God in space this way is to reduce God into an object, a thing among other things, a male patriarch among other humans. This objectification of Being was a major step in detaching, reifying, and weakening the presence and power of the Christian divinity.

The development of perspectival art in the Renaissance was a basic step in the direction of creating a mental construct of the objective world's space. Perspective is a system of pictorial conventions laid upon the world like a map's grid with vanishing points. But perspectival visualization, whether in drawing or photography, is not the generally assumed accurate picture of nature. As E. H. Gombrich stresses in *Art and Illusion*, "Perspective is merely a convention and does not represent the world as it looks" (254). Perspectival art expresses a worldview, a grid for oversimplifying and distorting the natural world into a new mental construct. Perspective especially aids in the scientific reduction of space to a static, homogeneous, rational, but audaciously abstracted framework. The retina would project in a spherical form rather than a pyramidal perspective form, and any art or photo on a flat surface reinforces this distortion by printing a segment of a spherical image on a flat surface (Panofsky 260). This reduces the full reality of space to the map of a geometric convention and suppresses its spirituality.

Perspectival visualization not only distorts the world; it also elevates the individual viewer to greatest importance, in the style of Renaissance individualism (Arnheim 294). The single vanishing point in central perspective is a visual metaphor for the singleness of the individual psyche enshrined as authoritative in the geometric schema created by perspectival art. On the horizon of perspective's program, the vanishing point collapses the infinite depth of Being into the subjective psyche of an individual with a "point of view."

By structuring the exterior world geometrically, perspective also paved the way for the positioning of things as if they were soulless objects "out there" in a mathematically reduced space, cleared of all desires. The deep mysterious soul of the world is swallowed up into the little dark room of perspective's viewer and into the selectively reductive construct of the object in geometric space. Stone lions in front of buildings, for example, lose their historical sense as guardians of a fortress meant to frighten attackers and show the strength of the monarch within. They become merely disenchanted, decorative stone "things" in a geometric world.

Like the shift in imagining time brought by the clock, the shift in art from symbolic to spatial portrayal during the Renaissance was a major building block that helped to envision the scientific world of homogenous space, where no unit of space is more sacred or holy than another. This space/time construct is not a self-evidently true picture of nature but a cultural construct with its enchanting vision of precision and the measurable, geometric quantities that make technology's powers possible. This construct also enchants the ego into imagining itself able to control the world it created by suppressing the sacred and elevating the quantitative coordinates of geometry.

One way that industrial culture does value one space over another is by judging a space's interest to the ego. For example, people out in unknown territory far from a town tend to say that they are "in the middle of nowhere." The ego grants value to towns, cities, and technologically full spaces and considers raw nature to be "nowhere." But a different ontology would see and honor the beauty and uniqueness of each space, whether mountain trail or desert cliff. Outside of homogeneous space, there is no such place as "the middle of nowhere" because each space has qualities to be honored, as John Muir discovered in Yosemite.

Fragments of World Become Objects

What we call "objects" are technological society's pictures of small fragments of the fullness of the world. "Subjects" stand outside and against objects. This stance also allows humans to imagine that we stand in value and power over the merely obtuse things that we see "out there." Thus we can dominate them. This philosophy is not only a well-developed technique built on the thinking of rationalist philosophers such as Newton and Descartes. There is also a trance, an enchantment built of unconscious desires underneath its thinking.

This way of imagining things as if they were nothing but obtuse objects is technological culture's narrow way of working in the world. But in itself there is no such thing as an "object out there." The objective evaluation of the world is not a matter of fact—it is a theoretical creation of the mechanical worldview. In its trance we select quantitatively measurable fragments of the world—resources and technologies—and allow them to serve as reality for human ends.

Quantification seems to make objects more certainly true constructs in time and space. Numerical time is a mechanical construct using clocks to coordinate machines, but it cannot capture the soul's qualitative experience of time. Space, abstracted by the Cartesian coordinates of three-dimensional graphs ignores the fact that space is shaped by bioregional variations and space and time warped by gravitational fields. When we think of the past and the future through the lens of this time/space construct, we are neglecting our collective mental role of projecting a mathematical system onto the time/space continuum. While time, space, and things are seen as only "out there," we know them through human creations. What we can know of the world of objects is seen through artificial systems and instruments created by subjects. Objectivity is a paradigm, attitude, and model for thinking—a pragmatic, utilitarian worldview created for the purposes of industrial culture. All its facts are thus theory-laden. Following Kant, B. Alan Wallace, in *The Taboo of Subjectivity*, says: "Every fact we can state about that world incorporates elements of our own thinking. The world that science studies cannot therefore be that thing-in-itself" (216).

Even though objectness is an incomplete paradigm, it is like a magical mirror that reflexively leads us to imagine ourselves ironically not as the dominating controllers that we think we are but actually as no better than objects. It reduces our sense of humanity and painfully stifles our souls, alienating us from the rich fullness of existence. Ira Chernus says:

> As we surrender to the machine, seeing our only meaning as servants who must keep the machine running smoothly, we merge into the machine and lose the sense of being truly alive. The ultimate end of this process might be called the total "thingification" of reality—the reduction of all life to meaningless inert objects. (146)

In certain objective methods of studying humanity, the subject itself ends up like a robot, reduced to a thing analyzed by the same methods as the object. At worst, humans are thus imagined as objects, numbers, and machines. Objectness attempts to stuff all the soul's participation in nature's depths and its passionate enchantments into the skull of the subject, striving to transfer its enchantments to another category named "subjectivity." The obtuse object is not a fact but part of a mechanistic paradigm, actually an imaginative *choice*, selected for human command, one of the most powerful values driving technological culture. And we let it reduce us to its level.

This culturally constructed world of supposedly neutral time, space, and objects is carefully constructed not only with the mental tools of reason, logic, and math but with the additional enchantments of the human will to power and pleasure. This constructed world is a powerful mystique that does not favor human harmony, women's rights, ecological balance, or animal rights but human, largely aggressive dominion. As the main ingredient, the obtuse object is a remarkably successful enchantment. But it is not necessary and not inevitable. The world of technique's objects trails clouds of hidden imagination that dissolve its certainty.

The world is not as objective as its conscious metaphysics tells us it is. Its underlying goals, purposes, and obsessions drive us to live in a world that seems to erase its color. Technologies embody desires for speed, and their utopian dreams lead to wild expectations with dark shadows. Space travel offers cosmic horizons, and robots seem like objects out there when they actually are mythic images of the mechanical metaphysics that we seem fated to live out.

Our great irony is that, in search of maximum control, we are losing control of the project of high-technology civilization. By treating the world as a set of dumb objects that we can control for our own unlimited power and pleasure, we are losing power, because this enchantment, which demands the deadening of the world, is turning back upon us. We damage our own nest as we rush headlong toward an imaginary mechanical utopia, regularly surprised by the unexpected destructive side-effects.

Objectivity's Nihilistic Wounds

A major distressing side-effect of the framework of objectivity is nihilism. If the only real, true world is the realm of objects, then where in it do the subjects fit and relate to objects? In theory, their rationality interacts with the objects, which are assumed to be knowable by the subject's objectivity, and the subject's passions are theoretically excluded as unreliable guides to truth. Ethics are reduced to rules, social contracts, or utilitarian calculations about pleasure or happiness that often fail to motivate people. And objectivity is too frequently guided by a subjective will to power and pleasure. So when human existential issues arise, such as questions about life after death, the spiritual beliefs are denigrated and feelings are repressed, except where other subjective human values have been smuggled in. Objectively, the universe is seen as cold, alien, meaningless, and even hostile to humans, so shallow quests for power and pleasure seem meaningful but too often cultivate cynicism. When you die, your subjective/objective existence evaporates into nothing. This metaphysic casts an unsatisfactory dark shadow on humanity and nature and inflicts a collective psychic wound called "nihilism," a gnawing sense that nothing is meaningful.

Subjectivity also points to a solipsistic nihilism. The isolation of the subjective ego in Descartes, Locke, Hume, and Sartre, for example, has left the subject grasping for "intersubjective objectivity," or ways of affirming meaning in life beyond the isolated subject. But the very structure of the subject standing against the object leaves little room to squeeze in meaning and purpose beyond crude survival or self-interest, leading to selfish relativism and absurdity. Thus the philosopher Donald Crosby, in his chapter on "The Subjectivist Turn," says: "What we assuredly know, therefore, is the contents and constructions of our own minds; all else is shrouded in darkness" (246). Crosby claims that it is nonsensical for an individual subject to wish that life had meaning or to arbitrarily confer meaning upon it (262). This is the nihilism of subjectivity, well expressed in Jean-Paul Sartre's 1945 play *No Exit,* in which he depicts the annoyances of other people as hell. This narrows the soul down to the pervasive stress on crude motives, selfish absurdity, meaninglessness, and the anguished nihilism of modernity—crude motorcycle gangs, greedy corporate scandals, and nuclear fireballs.

Objectively, it would not matter if the universe feels cold and alien, because these would be irrelevant subjective reactions to the bare factual truth. It may seem that this philosophy would give rise to a species of unfeeling, calculating people satisfied with a totally objective worldview. The mythic stone-faced cowboy, or the steely technological hero using a powerful machine, expresses this hardness. This severity requires a numbness that has been documented by writers such as Robert Lifton and Joanna Macy as a defensive response to the despair and terror of nuclear nightmares. Numbness emerges like a scab on a wound to cover pain and grief, but it then becomes a style of consciousness in the world of objectivity that suppresses authentic feeling. In the former Soviet Union, where cancer epidemics surround nuclear facilities, radiation leaks are obvious, and heroin addicts have been found operating nuclear reactors, an underfunded nuclear industry leader helplessly says, "'We must do our best and cross our fingers'" (qtd. in Nordland 40). The natural response to the nuclear nightmare that industrial society has created would be horror, outrage, and protest. Instead, there is resigned desperation and numbness. Similarly, Elisabeth Kübler-Ross has revealed the numb denial that is a commonplace defense against the overwhelming grief of death. Though grief's denial and numbness are receding with the development of programs such as grief counseling and Hospicare, they remain scabs on technological culture's painful nihilism. The archetypal psychologist Robert Sardello says, "[T]he physical world, the body of imagination, the world's body has entered a state of almost irretrievable numbness. No healing will be found for this numbness that will not entail a reversal of imagining the world" (35). Healing the nihilism of objectivity requires a major ontological reversal, not a little cultural tinkering.

The worldview of objectivity breeds nihilism, a pervasive numbness, denial, depression, cynicism, and despair. Some people hopefully compromise pure objectivity with some form of humanism, ethics, religious belief, or spiritual faith. Strong nihilism can be the result of what seems to be an optimistic objectivity but really has a strong subtext: the will to power.

The dominion of the will to power and objectivity required that God be swept aside, leaving the universe pictured as an infinite, meaningless void. This bold stroke opened the door to nihilistic wounding. Nietzsche's famous cry of the madman echoes through the wars, holocausts, and gulags of the twentieth century:

> "Whither is God?" he cried. "I shall tell you! *We have killed him,*—you and I. We are all his murderers! But how have we done this? How were we able to drink up the sea? Who gave us the sponge to wipe away the entire horizon? What did we do when we unchained this earth from its sun? . . . Are we not straying as through an infinite nothing? Do we not feel the breath of empty space?" (Nietzsche, *Gay Science*, 95–96)

The emptiness of space has turned out to be colder and more dangerous than when we westerners thought we had wiped away the entire horizon of traditional faith and cast away the caring God, distorted as it may have become. Certain elements of traditional faiths certainly need cleansing, but living in a postulated universe without care can leave one like an abandoned child, feeling horribly insecure, depressed, cynically angry, and even violent. Cosmic nihilism results when one grants totality to technical knowledge and then recognizes only blind cosmic forces originating in sheer chance, interpreting human existence as a tiny random accident in a vastly meaningless universe. This is not science but metaphysical enchantment. Life's sufferings and disappointments merely lead to lonely absurdity, moral emptiness, and death.

A major sign of living in a disenchanted world has been the persistent wounding from the denial of suppressed realities. There is a large literature of technological wounding. The romantics protested against excessive rationalism and technology, such as noisy, smoky trains, that pushed aside aesthetic and spiritual experiences. Contemporary critics of technology see a totalitarian worldview that indoctrinates industrial society's citizens with mechanistic, materialistic, and consumerist mentalities. T. S. Eliot and Theodore Roszak speak of the industrial "waste land," a hollow, meaningless, and absurd worldview that inflicts massive spiritual malaise: "A heap of broken images, where the sun beats, / And the dead tree gives no shelter, the cricket no relief" (Eliot 2583). Some modern art expresses chaos and grief, from Jackson Pollack's splatter paintings to junk sculpture. Atonal modern music evokes painful imbalance and confusion. The art of nihilism has even led some artists to videotape themselves bleeding on the floor (Perlmutter). Some psychologists conclude that industrial materialism damages self-esteem, happiness, and family and promotes anxiety, personality disorder, narcissism, and antisocial behavior (Kasser 22).

Excessive youthful suicides, violence, and alcohol and drug abuse point to a deep despair unhealed by technological culture. Developing countries around the world suffer from the stubborn refusal of industrial cultures to share their wealth. Magazines such as *Popular Science* gleefully promote "Gadget Fetish," the lust for the latest in high-tech gimmicks (Miller, "Gadget"), while peasants in other countries pick over garbage heaps for survival. Objectivity's cynical, nihilistic rush for power and profit amplifies and then ignores these problems.

Modern nihilism can be seen as a consequence of the forgetfulness of Being, the caring ultimate reality behind the personified, theistic God. Spiritual nihilism grows like a cancer from the emptiness left when ontological nihilism dismisses a caring ground of existence. It may begin as a fringe effect, but it spreads. Both technological culture's forceful denial of the reality of Being and its opposite, the angry defense of a fundamentalist version of Being, disclose the pathology of the forgetfulness of Being. A rigorously objectivist metaphysic cannot stand without igniting such side-effects, unexpected threats to the benefits of modernity. The forgetfulness of Being is a built-in, painful, and menacing enchantment of technological culture.

Digging deeper, we find that it is not only a fringe side-effect that nihilism emerges from the metaphysics of objectivity. The desire to set up the worldview of subjects standing against objects is a product of the will to power. The desire to dominate through human reason is the force that establishes objectivity and its inherent alienation and nihilism. As the philosopher David Levin says:

> [T]he imposition of this metaphysics of objects is the work of the will to power,
> which thrives on setting up situations structured in terms of the relationship

between a subject and its object. As the etymology of the word *object* (ob-ject, *Gegenstand*) tells us, this relationship is inherently organized around opposition, conflict, struggle, and violence. . . . [O]ur age has consequently become a time of terrible strife and destruction. (Levin 120)

This realization takes us a step further into the enchantment of technology, for it argues that objectivity's nihilism is not purely a by-product of rational analysis but is driven by the barely concealed enchantment with control and dominion. It is not the case that independently grounded rational objectivity is disturbed by an unruly excess of external, subjective nihilism. Rather, the unconscious, collective desire for control leading to nihilism uses the rationality of objectivity to achieve its purposes. Objectivity and nihilism are like Siamese twins, one optimistically refined and the other darkly crude. But they are inseparable, joined at the hip. One designs the machines, while the other angrily blows up the same machines and their designers in despair.

3   The Bottomless Subject

> The "subject" is not something given, it is something added and invented and projected behind what there is.
>   —Friedrich Nietzsche, *The Will to Power*

The Great Divide

"I celebrate myself, and sing myself, / And what I assume you shall assume, / For every atom belonging to me as good belongs to you," wrote Walt Whitman, intimating the underlying union of all selves (Whitman 852). Such poetry is merely trivial subjective whimsey, however, to the mechanical mind. Industrial society's basic understanding of the subject is limited to the internal, biased, opinionated, moody, introspective, individual feelings, sentiments, and preferences that exist only in a person's mind, unavailable for external verification. Science tries to push aside such subjective feelings to discover objective truth. In this metaphysical framework, the subject incorporates a range of experience radically differentiated from the object. The subject is internal, the object external. Descartes first distinguished the private thinking substance, the *res cogitans,* from the *res extensa,* or extended things. The world had not previously been seen in this dualistic way. Kant later began using the terms "subject" and "object" in this new way. One of the most deeply entrenched enchantments of technological culture is this theory that the mind is contained in a subjective mental region of the brain, looking out onto a separate world of objects. Living in this simplistic dualism is a central premise of what it means to be modern, but this notion is actually bottomless, for the modern subject is not master of its decisions. It does not know itself entirely, it is not fully "internal," nor is it "present" to itself, for unconscious passions lie below its well-lit control room and send powerful forces up to disrupt and guide it. Many of the "unexpected" outcomes of technology result from these dark, concealed forces, as we see in road rage.

Before the ascendancy of modern technology, the subject meant other things, such as a person subject to a king's rule, a grammatical category, or the essence of a thing. Modernity changed all that with Descartes and Kant's redefinition of the world into a two-sided structure straddling a great gap. Bruno Latour calls these " two radically different ontological zones, with human beings on the one side and nonhumans on the other, the 'Great Divide'" (10–11). Developed in this way, subjectivism has led to a powerful new sense of truth as objective, purged of subjectivity, but also to an excessive individualism and alienation from nature and Being.

The word "subject" has three meanings: first, the larger *interior* mental realm as distinct from the external physical world; second, the internal realm of desires, instincts, feelings, personal biases, prejudices, and values, distinct from reason; and third, the "objective" portion of the subject, known psychologically as the ego, which is metaphysically presumed to correspond to the outer world of objects, receiving perceptions from them accurately and rationally guiding interactions with them.

A major task of the rational part of the subjective mind is to separate the ego's objective rational, neutral, instrumental, factual, and scientific analysis of the world from the subjective internal mind, which is biased, distorted, and easily lured into superstitions and illusions by feelings that do not correspond to objective facts. But the subject's interior realm is full of contradictory tensions between desire and rationality. Technological culture generally grants priority to reason and relegates desire to a secondary level of truth and existence. Yet unconsciously, reason is bottomless. It drops out into a vast influential collective ocean of desires and enchantments.

Tangled Roots

The depths of the subject/object divide are rarely contemplated. The concept of the subject has tangled roots in ancient concepts. The Christian emphasis on individual salvation of the immortal soul helped draw the individual out of the tribal collective consciousness that was so common and strong in the ancient world. Christian belief in individual immortality is rooted in Egyptian, Persian Zoroastrian, and Platonic beliefs. It was developed by Augustine's *Confessions,* in which he examines his private life and his conversion to Christianity. The meaning of the word "subject" is rooted in the Latin translation of Aristotle's use of the Greek term for "substance," or the "essence" of mental, spiritual, or physical aspects of the world. Aristotle's Greek term was *to hypokeime-non,* "the underlying," from the verb *hypokeimai,* meaning "to lie under, to be put under the eyes, to be laid down, taken for granted, to submit, to be proposed or at hand."

This Greek word was translated into Latin as *subiectum*, which had three meanings: the underlying material or substratum out of which things are made; the subject of attributes; and the subject of predicates. The first definition is close to what we mean by the atomic structure of matter, although for Aristotle it also included the mental and spiritual realms. The second definition now applies to both "subject" and "object," and the third continues in the linguistic structure of some languages. This ancient usage was quite different from the modern Great Divide that separates the world into two grand categories, subject and object.

A different major archaic political meaning of "subject" is the verb "to submit." In this meaning, the person is subordinate, or subject, to a reigning monarch or a government and its laws, which reflects the Latin root of the word, "to throw or place under." Shakespeare expressed this in 1593 in *Henry VI:* "Was never Subject longed to be a King, / As I do long and wish to be a Subject" (part 2, IV.ix.5–6). Today, in more democratic societies, this meaning is implied in the "subject" of an experiment, as one who submits to and is the focus of study.

The Enlightenment project reversed this meaning of submission to a monarch to grant enfranchisement, human rights, and freedom to the individual subject. Today, the word maintains this sense of individual freedom of thought for democracies. The modern Cartesian and Kantian metaphysical definition has evolved into meaning the carrier of rational knowledge and desires and the isolated mind separate from the environment of objects. The meaning of the word "subject" has shifted from the ancient meaning of a physical and spiritual underlying carrier of attributes to the modern sense of a contained entity with emotional desires, rational knowledge, and political rights, standing opposite an objective world. The subject now carries the new implication of the inferior, merely personal, biased opinions rather than superior "objective facts." Among these shifting meanings, the validity of the latest cannot be taken for granted as self-evident. Postmodern thinkers have been deconstructing its meaning again. Jacques Derrida says, "There has never been The Subject for anyone. . . . The Subject is a fable" (Derrida 102).

The modern metaphysics of subjectivity has had a powerful effect on modern politics and psychology. Politically, the principles of human rights, which have taken root in some countries, have aided in the theoretical rejection of totalitarianism and dogmatism. Those who benefit can be grateful for this Enlightenment insistence on the right of the individual to think, speak, and believe freely. But this is a spiritual, ethical, and political platform. Individual freedom is not dependent upon the radical metaphysical disjunction with the "other" in objects. The subject is a constructed vision of personhood tied to the different project of conquering nature. To gain more power over nature, moderns had to adopt a radically new personality. As Morris Berman says, "[O]ur modern view of reality was purchased at a fantastic price. For what was ultimately created by the shift from animism to mechanism was not merely a new science, but a new personality to go with it" (Berman, *Reenchantment*, 113). Some of the negative effects of the heightening of subjectivity on the modern personality have been disengagement, isolation, loneliness, and alienation from community and nature. Yi-Fu Tuan has studied individualism in arenas such as households, manners, and theater, and he concludes:

> Western culture encourages an intense awareness of self and, compared with other cultures, an exaggerated belief in the power and value of the individual. The rewards of such awareness and belief are many, including the sense of independence, or an untrammeled freedom. . . . The obverse is isolation, loneliness, a sense of disengagement. . . . This isolated, critical and self-conscious individual is a cultural artifact. (139)

This culturally isolated individual construct is expressed in the cowboy riding alone into the sunset, the individual driving a car, and the astronaut floating in space. In *Habits of the Heart: Individualism and Commitment in American Life*, Robert Bellah et al. see this as highly ambivalent: "American cultural traditions define personality, achievement, and the purpose of human life in ways that leave the individual suspended in glorious, but terrifying, isolation" (6). Freedom to choose one's "lifestyle" and "values" became part of the linguistic currency of a fairly tolerant society, but this leaves a large shadow of lonely, isolated, depressed single people. The sense of belonging to a traditional community, with its authoritarian, oppressive, or narrow qualities, is left behind for another extreme: the freedom and loneliness of modernism.

This subjectivist disengagement also results in an alienation from nature that prevents respect and appreciation for the ecological wholeness of our home planet. Left behind is the sense of wonder at the awesomeness of our cosmic niche that inspires ecological ethics. We may enjoy the view of nature, the vacation, the river rafting, or rock climbing, but until we feel fully immersed in the world, consciously we still feel like isolated subjects in a world of alien objects. This is a high price to pay for the benefits of a high-tech society.

The Disembodied Soul

Part of the deep, imaginative background of this dualism is the ancient notion of the disembodied soul, the eternal soul that continues on after death and may be resurrected in the same body or reincarnated in a new body. This archaic religious vision of the disembodied soul was evident, for example, in the ancient Egyptian practice of mummification, apparently in the belief that the immortal soul would depart to the heavens. Plato added his picture of the disembodied immortal soul that was grafted onto Christianity later. These are some of the many archaic myths of the disembodied soul's journey. Such accounts of out-of-body travel, also described by contemporary survivors of near-death experiences (see Bailey and Yates), have a powerful spiritual significance, as part of Being, that in no way would support the earthly subjectivity that diverged from this origin by rejecting transcendence. Nonsubjectivist disembodiment in immortality envisions the soul returning to its heavenly origin. Industrial society's subjectivity theoretically leaves the isolated, mortal individual alone, standing against a neutral world of objects, and disappearing at death.

Ancient tribal shamans frequently undertook visionary travels into the transcendent realm of gods and goddesses to retrieve wisdom and treasures to help the tribe. Often in these visions they say they could fly above the earth and observe it as if detached. The Sioux shaman Black Elk is a classic example (see Black Elk chap. 3). Legends of the disembodied soul's journey may be one mythic basis for scientific detachment. It is an ancient tradition. Through the long process of shifting the meaning of this disembodied soul—through religions, Plato, Aristotle, and Descartes—the disembodied immortal soul slowly lost its religious and cosmic dimension. Instead, it became a more earthly, technical "mind" and developed its rational, logical, and calculating potential. By the time of Francis Bacon in the seventeenth century, the soul had become the subjective ego seeking dominion and power over nature and other people. Bacon made the early distinction between facts (natural knowledge) and values (moral knowledge) that strengthened the detachment of technological thinking from its context.

This dualistic philosophy is a partial picture of reality, a fragmented enchantment with many veiled difficulties. This is illustrated by Descartes's effort to exclude all sources of delusion and reach certainty in thought alone, when he said in his First Meditation:

> I will suppose that sky, air, earth, colours, shapes, sounds, and all external objects are mere delusive dreams, by means of which [an evil spirit] lays snares

for my credulity. I will consider myself as having no hands, no eyes, no flesh, no blood, no senses, but just having a false belief that I have all these things. (Descartes, "Principles," 65)

Descartes imaginatively discards all the environmental and embodied physical necessities that support his brain in order to ascertain supposedly certain truths deduced by a disembodied mind. This kind of disembodiment reduces thought to logical information alone, drastically eliminating a vast range of human experience. Descartes does invoke God to assure the certainty of his deductions, but God was soon discarded by others. Yet this enchanted disembodiment of the subject still echoes the archaic belief in the immortal soul with access to heaven. This detachment suggests an underlying enchantment with elevating the Cartesian mind to divine status, able to determine not only metaphysical certainty but moral certainty by detaching subject from object. Rejecting God, the Cartesian ego later assumes absolute powers.

This disembodiment is part of the Faustian Bargain—to gain unlimited power in exchange for disregarding contrary arguments about moral rightness—that has been made by technological culture. Such a bargain has been made with nuclear power, for example. Its advocates have argued that it is a Faustian Bargain from which we cannot turn back, one that is fundamental to our technological system. The strange fantasy that we cannot turn back from technological development is hardly an objective fact; it is rather a metaphysical commitment, a powerful enchantment that ignores the bodily remains left behind from this headlong voyage via the disembodied intellect in search of power. The immense problems of nuclear radiation on the environment and the body are tolerated in exchange for a vast supply of power.

A symbolic enchantment of the fragmentation of body and mind is expressed in the absurd disembodied "brain in a vat" fantasy. In the Frankenstein story, a brain can be attached to a body in a way that wildly imagines that it could exist independent of the nervous system and blood supply. Other versions imagine that an isolated brain could be somehow connected to a computer, as if such an organism actually shared the same communication system as machines. The questioning of the rightness of this image of the disembodied interior subject has been heated, especially since the development of Continental phenomenology and French deconstructionism. Bruno Latour says: "Why in the first place did we ever need the idea of an *outside world* looked at through a gaze from the very uncomfortable observation point of a mind-in-a-vat?" ("Never" 12). Washington Irving's classic story of the Headless Horseman evokes some revealing images of the disembodied intellect.

This Enchanted Head

Washington Irving wrote "The Legend of Sleepy Hollow," a yarn about the Hudson Valley's nineteenth-century Dutch descendants, in 1819–20. A gawky young schoolmaster, Ichabod Crane, is called to the task of formally educating the offspring of rustic farmers, who usually doubt the value of such endeavors. For all his education, however, he is not immune to the enchantments of the region. The farmers are held in a bewitching spell full of trances, visions, and an enduring reverie. They often see strange sights and hear music and voices in the many haunted spots that feed their supersitions. But the commanding vision of the region is the Headless Horseman.

When Crane's heart beats fast for the damsel Katrina Van Tassel, he angers another hopeful suitor, the local burly and mischievous Brom Bones. The competition begins, and Bones is determined to frighten Crane out of the region with an appearance of the legendary Headless Horseman, a spectral figure that haunts the countryside. This fearsome headless ghost is known for chasing lonely riders in the dark, throwing them into a brook, and then flying away with a clap of thunder. One dark night after a dance, Brom Bones, dressed as the headless horseman, Irving hints, meets Crane alone in the woods and terrifies him by chasing him until he and his saddle fall off his horse. Bones throws a pumpkinhead at Crane, who is sprawled on the ground. The hapless schoolteacher quickly disappears from Sleepy Hollow. This mortifying fate leads Crane to New York City, where he becomes a lawyer, while Brom Bones happily escorts Katrina to the altar.

The Headless Horseman holds subtle symbolic power. It can be seen as an image of the dualistic intellectual world that was emerging in Irving's time, which was only beginning to create the technologies that would follow from Descartes's philosophy that imagined a headless body and a bodiless head. This disembodied subject of Enlightenment rationality, separated from its natural body and its necessary environment, grew into an enchanted vision of technological culture less than two centuries later.

That same Headless Horseman still lurks in the dark shadows of contemporary consciousness, symbolizing the rationalistic goal of a disembodied consciousness, an intellectual tradition devoted to the enchantment of pure logic and theory—an immaculate subjective spirit torn recklessly from its objective body and environmental support by generations of Enlightenment thinkers. The remaining objective body without a head also suggests the theoretical world of objects, independent of any subjective thinker's ideas. That haunted, detached pumpkinhead flying through the air at the schoolmaster is the image of the modern subject itself.

Descartes Backward: "I Exist, Therefore I Think"

Perhaps the most pointed expression of the Headless Horseman's enchantment, the pithy quote that captures the thrust of the disembodied intellect, is Descartes's classic argument "*cogito, ergo sum*": I think, therefore I am. This was the outcome of his bold effort to eliminate all illusion. He began with the idea that true ideas are true by virtue of two principles: they are clear and distinct, not deceptive and confusing, like sensations or dreams, and their truth is assured by God, who would not deceive us: "[W]hatever we conceive very clearly and distinctly is true, is assured only because God is or exists and is a perfect being, and everything in us comes from him" (Descartes, "Discourse," 36). Descartes believed that many ideas are innate, such as the idea that there is only one God and that He is not deceptive. Descartes concluded that, no matter how many illusions he himself might entertain, even the illusion that he does not exist, he could nevertheless conclude that he does exist because of the bare fact of his thinking itself. Descartes's goal was to isolate pure reason from its distorting "subjective" illusions, found in dreams, imagination, and dogmatic religion. Then reason would be free to analyze the external world of objects (*res extensa*) objectively.

The implications of this enduring philosophical axiom are important. First, even though Descartes himself may not have intended it, since his argument's certainty requires a philosopher's logical God, the *cogito, ergo sum* formula has become an elevation of pure conscious reason to be the most important element of the mind and a devaluation of imagination, intuition, and religious experience. Second, it strives to purify clear and distinct thought sufficiently to guarantee its freedom from any "subjective" illusions, even though his argument requires imagining a disembodied mind. Third, although Descartes's argument was inspired by dreams, this purification effort simply pushes below the surface of conventional consciousness all elements involved in thought that are characterized as nonrational, creating what later came to be called the "unconscious" (Whyte 86–90). Fourth, it grants the logical subject (*res cogiants*) the authority to determine what is real. With God as the background guarantor, later dismissed by others but remaining at its unconscious roots, Descartes elevates the notion of the disembodied subject, separate from the external world, to be the authoritative determinant of clear and distinct truth, even though a clear and distinct statement could easily be false. Now the subject, not God, is supposed to be the guarantor of truth. On this, Martin Heidegger writes:

> Inasmuch as Descartes seeks this *subiectum* along the path previously marked out by metaphysics, he, thinking truth as certainty, finds the *ego cogito* to be

> that which presences as fixed and constant. In this way, the *ego sum* is transformed into the *subiectum*, i.e., the subject becomes self-consciousness. The subjectness of the subject is determined out of the sureness, the certainty of that consciousness. (Heidegger, *Question*, 83)

However, if we look beyond the isolated mind at the context of Descartes's proclamation, we find other determinative factors. How could this man think unless he had an education, teachers, books, past scholars to study, and a community of thinkers with whom to discuss his ideas? How could his brain operate without the support of the rest of his body? Have food, parents, warmth, blood, nerves, or perception nothing to do with his thinking? Descartes briefly acknowledges this but ignores it in his main argument ("Earth" 229; "Third Meditation" 80). Were there no unrecognized psychological supports to this thinking? No sense of security, family, love, courage, or adventure behind his mental explorations? Did he not exist because of the millennia-old biological system of *homo sapiens*—digestion, breathing, reproduction, and so on? How could he exist without the earth's environmental support for life? Existence or thinking can only be reduced to such a narrow foundation as pure reason by ignoring a vast range of experience. When subsequent thinkers dispensed with Descartes's guarantor of certain truth, God, yet kept his dualism, they lost the certainty that he claimed while still believing that they knew what was illusion. Nevertheless, this narrow foundation remained the basis of dominant understandings of certainty, nature, and technology.

Descartes's absolutist claim of certainty expresses the seventeenth century's royalist and Roman Catholic absolutism, which was embodied in princely despots, monopolistic financiers, and operatic prima donnas: "Descartes' soloism was a natural expression of baroque absolutism" (Mumford, *Pentagon*, 81). The reign of the separate individual that began in the Renaissance was now adopting a stance of intellectual absolutism, reflecting the way that its popes and kings claimed unconditional authority and absolute certainty. Descartes wrote in 1630: "'God sets up mathematical laws in nature, as a king sets up laws in his kingdom'" (qtd. in Berman, *Reenchantment*, 111). Descartes lived during a time of bloody wars over religion and the rise of the Protestant middle class in Europe. He saw the rise of the tyrannical Bourbon dynasty and the cynical manipulations of politicians such as Cardinal Richelieu behind the throne of the young Louis XIII. Like many thoughtful people, Descartes sought an intellectual foundation free of the dogmatic violence surrounding him. Yet his formula did not free itself from the surrounding enchantment of absolute certainty, whether dogmatic, individualist, or royalist. Instead, it repeated it in a new metaphysic.

The subsequent reverence for Descartes's godlike claim unconsciously carries forth this dream of absolute certainty, even though the political context of its origin has been forgotten. This claim to certainty is the enchantment of Descartes's mantra *cogito, ergo sum*. Thus he helped found the modern notion of the mind as the subject standing in opposition to a world of objects. However, modernism's pretense of certainty is weakened, once we see how its hidden mystique of absolutism replaces the concept of God. This formula has claimed objective certainty by promoting the disembodiment of the mind, removing it from its subjective unconscious depths, its bodily foundation, and its natural environment. Others later cast away its very life support: Being, formulated as a rationalist God but retaining its royalist claims. A better formula for those who want to reunite the painful Great Divide between subject and object would be Descartes's assertion reversed: "I exist, therefore I think."

Metaphor at Work

If we think in the lifeworld of existence rather than simply thinking "in the head," subject and object are bridged. For example, when San Francisco's Golden Gate Bridge was completed in 1937, the city basked in the civic pride it evoked. As David Nye reports in *American Technological Sublime,* the *San Francisco Examiner* saw it as

> "a gateway to the imagination . . . in its artful poise, slender there above the shimmering channel, it is more a state of the spirit than a fabricated road connection. It beckons us to dream and dare. First seen as an impossible dream, it became a moral regenerator in the 1930's for a nation devastated by depression" . . . "proof" . . . that the nation's "inventive and productive genius" would prevail. (xx)

One can still look at the bridge as an object, as its engineers rightly did when they built it, but should they feel no pride in their accomplishment? Should they have nothing to do with the celebratory opening rituals and blandly say, "The bridge is just an object; don't be subjective about it"? We tend to think of metaphors as verbal ornaments, subjective foam that is irrelevant to the "hard" facts of an objective world. But here is an object that is not merely functional but also a metaphor for communal pride and hope, a "moral regenerator." We can conceptually distinguish these feelings from the bridge as object, but in living experience they are unified phenomena. The bridge is a sublime icon of technology's transcendent significance.

Bridges can symbolize a connection between two realms. In the ancient Persian Zoroastrian religion, the Chinvat Bridge was traversed by deceased souls into the next life. Similarly, metaphors symbolically connect conscious-ness to unconscious soul. Metaphors work like bridge technology, but that work is collective, not personal or private, as the Golden Gate Bridge is a symbol of communal or collective hope and connection. Surely those bridge engineers could switch back and forth between being rational about their calculations and enjoying the city's pride and celebrations, letting the bridge become a metaphor as well as an object. We need to distinguish between imaginative metaphor and objective engineering techniques, and each may be brought to consciousness separately. But in daily life, the two intertwine in the soul. Bridge as steel is bridge as pride.

Similarly, automobiles personify the soul, blending person and machine. People casually say, "I am parked over there," "He skidded to a stop," or "She sure is revved up!" This may be called a process of personalization of a technol-ogy, and though it is playful, it indicates identification with one's car in a way that is quite different, for example, from riding a horse. When we think we are operating a car in a purely technical way, we may be quite involved in a narcissistic fashion, unconsciously letting the car be an expression of a trance of freedom, power, and control. Going over the Golden Gate Bridge in a car offers a sublime experience of freedom that transforms driving far over San Francisco Bay into a thrilling leap from land to wondrous flight far above the waves.

Machines shape personality, and personality uses machines to express itself. Machines have long offered this function. In *The Taboo of Subjectivity*, B. Alan Wallace says that in the midst of the scientific revolution,

> some natural philosophers likened the mind to a hydraulic system, and an early twentieth-century metaphor for the mind was a telephone switch board. . . . [S]cientific materialists have long been convinced that it must be similar to some kind of ingenious, material gadget. (126)

Mechanical metaphors go to work in expressions such as "turn me on," "pump it up," or "I am programmed." If we start with an image of the mind as a machine, then our conclusion is bound to include that same image. So the metaphor, meaning to be explanatory, becomes a determinative subjec-tive fantasy that guides subsequent thinking, as we are seeing today in the widespread image of the mind as a computer. We unconsciously merge soul with machine, saying, "the computer thinks," "the car is stupid," or "the robot is confused." None are literally true. This is speaking poetically, but we don't realize it. Metaphor is more than a disposable subjective icing on the cake of facts. It becomes a paradigm that shapes the very way we deal with the objective world of bridges and cars, by the way absorbing imagination into rationality.

The term "technological fix" promises a solution that may not develop, as with the first organ transplant in 1969, which was seen as great success, even though the patient died soon after the operation (Nelkin 42). Was this really a "fix"? Even if that historical event was seen to be on the path to a fix, the idea of a fix neglects nontechnical aspects, such as preventive community medicine, that are less dramatic public health issues.

Of course, we know better now, but in the 1940s lobotomies were praised as "no worse than removing a tooth" and seen as promising for "cutting out cares, relieving unbearable pain, checking feeblemindedness, curbing psychosis, helping schizophrenics, solving crime, and curing epilepsy" (Nelkin 48–49). Lobotomies, however, turned out to be costly and to have enormous variations in benefits. Above all, potential ethical abuses were sometimes ignored (48–49). This is hardly comparable to removing a tooth. Dorothy Nelkin stresses that we cannot avoid acknowledging "the judgmental and emotional factors that enter into scientific evaluation when it serves the needs of public policy" (61).

The problem is that worn-out metaphors still work hard as foundations of thought. "Is language the adequate expression of all realities?" asked Nietzsche, in his typically brassy way.

> And what therefore is truth? A mobile army of metaphors, metonymies, anthropomorphisms: in short a sum of human relations which became poetically and rhetorically intensified, metamorphosed, adored, and after long usage seems to a nation fixed, canonic, and binding: truths are illusions of which one has forgotten that they *are* illusions; worn-out metaphors which have become powerless to affect the senses. ("Truth" 506–8).

The subject/object dichotomy is a metaphor that was once fresh but now has become canonical and a stale basis for the industrial framework. Elaborated metaphors such as Frankenstein portray the surging currents beneath technological culture. Machines such as cars are built as metaphors themselves, or like the Golden Gate Bridge, they become powerful, influential premises. Language, seen as Nietzsche's "mobile army of metaphors," is rooted not only in objective facts but in poetic images that become fossilized but still actively shape consciousness.

New York's World Trade Center towers, targets of the tragic attack on September 11, 2001, attracted assault because they were an American icon. Their height, symbolizing massive technological power, pride, and wealth on Wall Street, was the idea of a public relations man, not the architect. As Cathleen McGuigan says, "Skyscrapers are an American invention, and the World Trade Center was among the last to reflect something of the visionary

ideals of techno-progress that so defined the last century" (McGuigan 87).
"How high can we build?" dreams industrial culture; "How high can we fly?"
The audacious hijackers who slammed two fully fueled airliners into these
metaphors knew well that the towers inspired monumental awe, fascination,
and resentment to many around the world. The towers symbolized to them
America's excessively hoarded wealth and blatant hubris. They might as well
have had targets painted on them. They exploded and collapsed into a monstrous tragedy not because they were utilitarian "objects" in a neutral Cartesian
space but because they were metaphors, the language of the proud collective
soul of U.S. capitalism. They displayed phallic pride, competition, and wealth,
as sublime public relations icons, tall enchanted cathedrals of techno-self-ag-
grandizement, rising in vulnerable steel.

Metaphor is involved in language from the beginning, not as an imaginative element added to purely factual perceptions. As Owen Barfield says:

> [T]he mind is never aware of an idea until the imagination has been at work
> on the bare material given by the senses, perceiving resemblance, that is, de-
> manding unity. We can go further than this: the mind can never even perceive
> an object, *as* an object, till the imagination has been at work combining the
> *disjecta membra* of unrelated percepts into that experienced unity which the
> word "object" denotes. (*Poetic* 25–26).

It would be inaccurate to say that metaphors contribute to the idea of subjectivism in any causal fashion. Images offer organizing patterns to conscious
data but cannot be said to *cause* ideas. Of course, reason has some autonomy,
but even the idea of cause and effect has its own root metaphors, such as the
great chain of being linking gods to earth. Such images govern ways of seeing
and establish orthodox ways of thinking. Such archetypal images sparkle with
their "emotional possessive effect, their bedazzlement of consciousness so
that it becomes blind to its own stance" (Hillman, *Re-visioning*, xiii). Guiding
images unconsciously organize patterns of understanding.

Historically, machines have become enchanting metaphors for organizing our relation to the universe. The wheel is a natural ancient image, given
the rotary movement of the earth that appears in the sky and earthly cycles.
The potter's wheel, the fiber spindle, the lathe, and Plato's cosmology of a
spherical heaven revolving around a cosmic axis of light all offer images for
organizing experience. The grand mechanical town clock in a tower offered
another lasting image for the universe. The mechanical clock's regularity,
uniform units of time, and controllable manufacture lent itself to the Deist
metaphor of God as the cosmic clockmaker. Descartes organized his views
with the root metaphor of the animal as a machine: "[I]f there were machines

with the organs and appearance of a monkey, or some other irrational animal, we should have no means of telling that they were not altogether of the same nature as those animals" ("Principles" 41). This audacious thought experiment assumes a reductive mechanical world rather than explaining it.

Post-Cartesians, who removed God from the system, came to emphasize control, and then mechanical thinking: "To place nature under our rule, we think of it more like a machine. The same thing with ourselves, as our internal state can be solved as problems only if we, too, work along rational principles" (Rothenberg 112). The entire scientific/technological method of reducing the world to its smallest components and rebuilding it artificially and the factory method of using identical parts is rooted in the clockwork-mechanism model of nature. In 1748, the French writer Julien Offray de La Mettrie, talking of springs and self-movement, simply proclaimed, "Man is a machine" (148). After the clock image wore thin, the metaphor of the cosmos as engine emerged with steam engines, and now the computer enchants many theories. Along the way, two machines quietly contributed to the image of a subjective mind as contained in the skull: the camera obscura and the magic lantern.

The Camera Obscura: Skull's Dark Room

There are many technologies that have quietly heightened our sense of being subjects in a world of objects. One is the camera obscura. This simple old machine is a dark room with a small hole that permits daylight to enter. When a white screen is set up opposite the hole, the external scene will be projected onto the screen upside-down. Place a lens in the hole and the picture can be inverted and focused. In the Middle Ages the camera obscura was used to view eclipses and to study light. This dark room slowly became an unconscious root metaphor for the growing philosophy of the subjective mind cut off from the outer objective world, communicating through the eye's tiny light channel. Now, whenever we sit in a dark room to watch a movie, a slide show, or television, we enter the imaginative world of the camera obscura, the dark room of the soul (Bailey, "Skull's").

In 1839, Louis-Jacques-Mandé Daguerre inserted photosensitive plates into a camera obscura and invented the photographic camera. The camera obscura began as an experimental model for the eye, was used by artists to create more realistic painting, and grew into a ruling metaphor for the mind. Offering a way of picturing the Cartesian internal *cogito* with a sensory channel admitting pictures from the external *extensio*, the image of skull's dark room developed. Over time, this image grew from a suggestive experimental analogy into a concealed metaphysical paradigm.

The philosopher Sarah Kofman highlights the impact of the camera obscura in *Camera Obscura of Ideology*. She demonstrates how the root metaphor of the dark room receiving projections inward has been used to support two contradictory western ideologies. Leonardo da Vinci and Jean-Jacques Rousseau argued that the representation of the external world inside the camera obscura supports the view that the pictures in the blank screen of the mind *accurately* represent objective reality (representationalism). But conversely, Marx, Nietzsche, and Freud saw the image of psyche modeled on the dark room as evidence that unreliable *delusions* are projected into the blank screen of the mind's inner chamber. To them, the images shining on the mental screen are delusory inversions of social relations and knowledge, the origin of which has been forgotten. The question becomes whether to trust or distrust the incoming sensations.

More recently, thinkers have come to see the camera obscura as picturing psyche isolated in a subjective container that has become a mental prison. Owen Barfield explores this in *The Rediscovery of Meaning*. He reminds us that the "onlooker" stance is not natural. Rather, mankind has had to "wrestle his subjectivity out of the world of his experience by polarizing that world gradually into a duality. And this is the duality of subjective-objective, or outer-inner, which now seems so fundamental" (16–17). According to Barfield, the camera obscura was instrumental in bringing this mental construction about. The Renaissance *chambre obscure* and its descendant, the nineteenth-century photographic camera, have inaugurated the age of "Camera Man." The camera obscura has long offered an imaginative structure leading to the vision of the soul shut up in a dark box, peering out at the world from a stance of isolated, lonely, alienated, subjective egos. For Barfield, the age of Camera Man is not an inevitable development but a distorting aberration.

The Artificial Eye of the Soul

Leonardo da Vinci made the earliest recorded explicit comparison of the camera obscura to the eye, calling it an artificial eye:

> "When the images of illuminated bodies pass through a small round hole into a very dark room, if you receive them on a piece of white paper placed vertically in the room at some distance from the aperture, you will see on the paper all those bodies in their natural shapes and colors, but they will appear upside down and smaller. . . . The same happens inside the pupil." ("Manuscript D" qtd. in Gernsheim and Gernsheim 19)

Da Vinci sketched the earliest surviving drawing of a camera obscura about 1508, "to bring the image of a crucifix into a room" (DaVinci I:133–34). He envisioned accurate representation of external images on the internal screen, and so he compared the camera obscura to the eye, trusting the incoming sensations. By the mid-sixteenth century the camera obscura was known in England, France, Holland, Germany, and Italy (Gernsheim and Gernsheim 18–19).

By Descartes's time the artificial eye was considered an important test of the reliability of sensations. He discussed an experiment in *La Dioptrique* in 1637:

> If a room is quite shut up apart from a single hole, and a glass lens is put in front of the hole, and behind that, some distance away, a white cloth, then the light coming from external objects forms images on the cloth. Now it is said that this room represents the eye: the hole, the pupil; the lens, the crystalline humor—or rather, all the refracting parts of the eye; and the cloth, the lining membrane, composed of optic nerve-endings. ("Dioptrics" 245)

Fitting an ox's eyeball into the room's opening to demonstrate the eye–dark room analogy, Descartes discovered the wonder: "You will see (I dare say with surprise and pleasure) a picture representing in natural perspective all the objects outside" (245). He was forced by the camera obscura to admit the accuracy of the representations of external bodies on the blank screen. John Locke would later take this to indicate the reliability of visual perception. But Descartes insisted that despite their apparent reliability, the images on the screen do not overcome the problem of delusion, which could be located further back in the brain's passages. The camera obscura is not a neutral tool revealing a supposedly objective truth about perception or epistemology. It is an enchanting metaphor, map, or model for a way of imagining the psyche as if it were held in a container. Don Ihde rightly calls the camera obscura an "epistemology engine" because it models the epistemology of representation so well, suggesting an accurate interior mental correspondence, free of interpretation, with the external world in the process of knowing (Ihde, "Epistemology").

Locke expanded the analogy of the camera obscura to let it represent the whole of human understanding. In contrast to Descartes, he believed the interior images to be trustworthy representations of the external world. Because of his faith, Locke awarded this dark closet the honor of becoming a key metaphor for human understanding. In his *Essay Concerning Human Understanding* (1690), he modestly proclaimed:

> *Dark Room.*—I pretend not to teach, but to inquire, and therefore cannot but confess here again, that external and internal sensation are the only passages that I can find of knowledge into the understanding. These alone, as far as I can discover, are the windows by which light is let into this dark room: for

methinks the understanding is not much unlike a closet wholly shut from light, with only some little opening left, to let in external visible resemblances, or ideas of things without: would the pictures coming into such a dark room but stay there, and lie so orderly as to be found upon occasion, it would very much resemble the understanding of a man, in reference to all objects of sight and the ideas of them. (Book 2, chap. 6, pt. 17)

Locke elevates the image of the blank screen to imaginatively symbolize his view of the understanding as a *tabula rasa*, an empty tablet that needs no innate principles to explain knowledge. In contrast with Plato's view of innate ideas such as justice, Locke proposed that the understanding is completely determined by sensory perceptions etched onto the mind's blank screen that receives these representations of the world.

Since Locke's representationalism formed the foundation for much of western psychology, the camera obscura enchantment entered into western thinking as a classic yet largely unacknowledged root metaphor for the psyche itself, as if it were nothing but a subjective dark room. Every blank screen—cinematic, television, or computer—still replicates this metaphor.

The Private Eye of the Soul

For artists such as Da Vinci who used a camera obscura, the device became a personal, individual experience, a private, contained box for peering out into the world. The dark room of the soul became a solitary event. Small, portable, and even hidden, camerae obscurae became instruments of private consciousness. Da Vinci's note is the earliest surviving description of a portable dark room:

> A cubic box should be made of wood with its sides firmly fixed together, except for the one in front which has to be taken over by the plate of iron [with a small hole in it], and the opposite one at the back, which is made of a thin sheet of paper or parchment, glued around the edges of the wooden box. (DaVinci I:133)

This headlike box offered the image of the psyche contained in a box. Johan Kepler, making a survey in Austria in 1620, also used such a portable dark room (Gernsheim and Gernsheim 23–24).

The fantasy of having the psychic power to secretly observe others also pressed the camera obscura into service in the seventeenth century. Hidden *chambres obscures,* as the French spies called them, were disguised in many ways. They were hidden in drinking goblets, books, walking sticks, and carriages (Gernsheim and Gernsheim 25–28). Even today in cinematography, the dominant custom is for the camera to be treated as an unacknowledged, if not secret, observer. The concealed "candid" camera is the latest trick in this long line of images of the hidden, subjective psyche.

Portable, secret observation lets the viewer retreat into the little black box of subjectivity and observe, imagining an alien world "out there" in which he or she does not participate. It models the fantasy of a purely mechanical world of soulless objects populated with isolated, contained subjects who learn by receiving projections. Divided at the Private Eye of the lens, the portable *chambre obscure* separates the world into the external mechanism and the hidden ego. This enchanted box portrays the mystique of contained subjectivity.

The Projected Metaphysics of the Soul

During the Renaissance, the ghosts, devils, and demons of popular religious mythology, long believed to influence events from dark mysterious places, began to appear moving about in the new psyche-box. Itinerant performers used the camera obscura to frighten audiences in dark fairground booths. An exposé of their techniques was recorded in Belgium by Francois d'Aguillon in 1613:

> In just this way certain charlatans tend to hoodwink the uneducated rabble; they pretend to know about black magic, while they are hardly aware of what that means. They boast that they conjure up phantoms of the devil from hell itself and show them to the onlookers. They lead the inquisitive and curious, who want to know all about secret and obscure subjects, into a dark chamber where there is no light, except a little which filters through a small pane of glass [lens]. Then they tell them sharply not to make any noise and to be as quiet as a mouse. And when everything is completely still and no one either moves or says a word, as if they were waiting for a church service or a vision, they say that the devil will soon come. In the meantime, an assistant puts on a devil mask to make him look like the pictures of devils one usually sees, with a hideous, monstrous face and horns on the head, with wolf's pelt and tail, with claws on his sleeves and shoes.
>
> Then the assistant struts up and down outside [the camera] as if he were deep in thought, to the place where his colours and shape can be reflected through the glass pane into the chamber. To make these cunning intentions even more effective, one should remain absolutely still, as if a God were to arise by way of this artifice. Then a few begin to go pale; some, out of fear of what is to come, start to sweat. After that, one takes a large board of paper [a piece of cardboard] and stands it opposite the light-rays which have been allowed to enter the chamber. On it can be seen the picture of the devil walking up and down; this they look at with trepidation. This is the reason why the

poor and unexperienced are unaware that they see the charlatan's shadow and squander their money unnecessarily. (D'Aguillon, book 1, prop. 42)

The devil, commonly pictured in sermons, church art, and popular theater, now dared to appear moving about mysteriously in the camera obscura. While earlier static images were understood as conventions, the new, strangely moving images inside these dark rooms were terrifying. But as the fear of thunder and lightning is diminished by scientific explanations, fear of the devil can be decreased by reducing him to natural trickery. The critic who delighted in revealing the fairground trickery also helped to reduce visionary metaphysics to natural optics.

Exposés of shows like this eventually supported the argument that mythic creatures such as the devil are "nothing but projections." Once the actor outside the room is unveiled, the psychic event inside the room seems less fearful. Such demystification of imaginative figures implied the correlative assurance that "it's all in your head." The world's terror, guilt, and evil symbolized by the devil can now be imagined as merely the passions felt inside the darkroom of the soul. The emotion experienced in the camera obscura is relieved upon discovering the "objective facts." That relief can turn into iconoclasm, skepticism, and even a defense against the soul's passions. Slowly but surely, the inner self and interior feelings develop into the new locus for devils and angels alike. Eventually anyone believing in angels, devils, or God could be accused of foolishly "projecting" them out of their subjective mind into the objective world.

Similarly, when Sigmund Freud postulated that most mythology and religion "is nothing but psychology projected into the external world," he imagined the psyche as a contained, internal reservoir that erroneously flows out into the Newtonian machine-world (Freud, "Psychopathology," 259). Mythology belongs inside the head, he presumed. Like the dark booth at a Renaissance fair, Freud's type of psyche belongs inside the container. The only reality of the soul's experiences lies inside the psyche-box—the skull's dark room—as the enchantment of this root metaphor insists.

That Obscure Chamber of the Subject

The idea that the soul is restricted to an internal chamber led to the image of "encapsulated man," a sense of contained psyche that fragments knowledge. Joseph R. Royce uses this term to denote a Cartesian sense of contained certitude: a limited, unjustified, myopic sense of certainty based on inadequate, relative truths:

What do we mean by encapsulation? In general we mean claiming to have all of the truth when one only has part of it. We mean claiming to have truth without being sufficiently aware of the limitations of one's approach to truth. We mean looking at life partially, but issuing statements concerning the wholeness of living. In its most important sense the term "encapsulation" refers to projecting a knowledge of ultimate reality from the perceptual framework of a limited reality image. (30)

Royce's definition certainly applies to what we call "subjective" claims, including religious dogmas and objective claims to truth, such as the Soviet system of "scientific Marxism" or an overly mechanistic view of humanity as machines.

Lewis Mumford uses the phrase "encapsulated man" to refer to pathological solitary confinement, a mechanical return to the womb, an underdimensioned view of the soul captured, for example, in the astronauts' confinement to a tiny space capsule perched on top of an explosive rocket (Mumford, *Pentagon,* photo 14–15). The astronauts seem to be temporary mummies echoing the purpose of the Egyptian pyramidal megamachine: to propel the eternal soul of Osiris back to his starry home. This archaic belief in magical travel to heaven is re-created through powerful rockets and computers with astronauts protected from cold space in tiny capsules, but its purpose hides the same ancient desire: to explore the heavens like a god (306). Why cannot the same brilliant reasoning that made these rockets now see the enchantment behind the logic? The subjectivized "encapsulated man" poses the "danger of cutting off pure intelligence from all its self-regulating, self-protecting organic sources" (313).

Self-encapsulization is a dangerously restrictive distortion of the psyche's natural participation in a deeper soul and world. This subjectivism, externalized by the camera obscura and the astronaut's space capsule, adheres to the Newtonian metaphysical notion that the world is nothing but a vast collection of objects, organized into a machine, which has no place for the psyche except inside our dark skulls. To serve physics, subjectivism theoretically scrubbed the world clean of ancestor spirits in the lakes and gods in the thunderclaps, clearing the way for the development of technology. But this cleansing represses our awareness of the irrepressible sense of deeper, mysterious wonders in the universe, and this suppression has become seriously problematic. As a paradigm for isolating subjectivism, the camera obscura needs to be demystified so the soul can break out of its long imprisonment in the obscure chamber of the skull's dark room. In addition to receiving projections inwardly, the subjective psyche is also imagined to project outwardly.

Skull's Lantern and "Projection"

Many an enchantment that comes out of the psyche's dark room is labeled a "projection." In modern psychology and religion, projection is the theory that we discover our unconscious contents out in the world, for example, when we fall in love at first sight, hate an enemy barely known, see God as a heavenly father, or, even philosophically, when we imagine the world to be a set of objects. In psychotherapy, we make many valuable discoveries about our own unconscious by examining and "owning" projections in quarrels, fantasies, and illusions. Begun by Feurbach, developed by Freud, and expanded by subsequent psychotherapists, this theory has opened many doors to the dimly perceived or concealed feelings lurking beyond consciousness.

But what unconscious presuppositions are disclosed when we ask, "What are we 'projecting' in the theory of projection?" The theory has been subjected to fruitful rethinking from various points of view (Holt; Von Franz; Fortmann; Bailey, "Skull's"), but the root metaphor of projection itself remains unexplored. Everyday familiarity with cameras and movies has veiled the influence of photographic projection on psychological theory. The idea of projection in psychology has its imaginative roots in the slide projector and its descendants, the movie and video projectors. The slide projector, known after its origins in the Renaissance as the "magic lantern," has quietly guided western ways of thinking about the nature of imagination, delusion, and subjectivity. The magic lantern exhibits the philosophical notion that a purely internal psyche projects subjective feeling outward onto a soulless world of objects.

Only rarely has the connection between the magic lantern and the idea of psychological projection been recognized. Carl Jung once remarked, following a dream of a magic lantern hovering above him like a UFO, "I am projected by the magic lantern as C. G. Jung" (Jung, *Memories*, 323). And Owen Barfield once briefly pointed to the image of the magic lantern hidden behind the western philosophy of subjectivity. He said that in the theory of projection, fantasies are seen

> as a sort of unconscious "projection" of the inner life of feeling and impulse upon an inanimate outer world. Now whenever this excellent word "projection" is spoken or thought, the humble student of meaning . . . makes his bow to an ingenious little machine . . . [the] "magic lantern." For the curious idea of "projection" which this invention begot in us, is now . . . ubiquitous. (Barfield, *Poetic*, 203)

When this inner subject/outer object dichotomy rose to prominence in western thought, Barfield says, it was imaginally strengthened by the magic lantern.

Western subjectivism has been shaped by the root metaphor of the magic lantern—*skull's lantern*—but we are only dimly aware of it.

The camera obscura worked its magic as a passive projection device, receiving moving images from the external world. The magic lantern was an active projector of artificial images painted on glass slides. As a hidden collective enchantment, the magic lantern has misled psychology into imagining that psyche actually projects outward from a contained mind. This image has built into the theory of psychological projection the pervasive distortion that the psyche is properly restricted to the skull. Thus, when experienced outside this container, psyche must always be a delusion, an error that should be "withdrawn" back inside.

Freud's use of projection as a psychoanalytic construct was new, but the idea of projection was already commonplace in late nineteenth-century European thought. When Ludwig Feuerbach argued in 1841 that "what by an earlier religion was regarded as objective is now regarded as subjective" (13), he presumed not only the subject/object dichotomy but also the hidden image of the camera obscura. The real experience of gods and devils occurs only inside the subjective dark room of the soul, Feuerbach contended, and is only erroneously externalized or projected outside it.

Friedrich Nietzsche criticized ideas of psychological projection as being excessively subjectivist:

> The fragment of outer world of which we are conscious is born after an effect from outside has impressed itself on us, and is subsequently projected as its "cause." . . . There is no question of "subject" and "object." (*Will* 265–67)

All this theorizing about projection took place in a nineteenth-century context of widespread traveling shows across Europe and America that used the magic lantern to enchant audiences. Crowds packed theaters to view the latest device for projecting mysterious ghosts, angels, or skeletons onto smoke screens, hidden glass panes, or translucent curtains. Popular magazines revealed these theatrical secrets. Long before the psychologists and philosophers began to speak of projection in psychology, these excited audiences were offered a way to think about the image-making activity of the subjective mind.

The disenchantment of the world was aided by the magic lantern because previous spiritual experiences, called "apparitions" or "ghosts," were now apparently reduced to a technical theater trick, the work of a clever lantern called "magic." The magic lantern helped remove "magical" phenomena from the external world to the interior, subjective world of the mind. The magic lantern's enchantment helped create the philosophical subject that attempts to interiorize soul into the skull.

The transformation of the slide projector into the motion picture projector began in the 1880s in Thomas Edison's lab and in France in Auguste Lumiere's workshop. Combined with rapid technical developments, projected moving images soon blossomed into the modern cinema industry. And all this imaginative activity stimulated more thinking about projection as a psychological phenomenon. Today, the camera and the projector work as a team: the camera passively records, and the slide projector, movie projector, television tube, or computer graphics program and data projector actively cast forth the recorded images. The viewer of any of these projections sits in a camera obscura, watching a descendant of the magic lantern, unaware, perhaps, of the influential theory that psyche projects its subjective contents out into the objective world like a magic lantern (Bailey, "Skull's").

"Nothing but Projections"

The word "projection" was adopted as an image of mental activity well before the Renaissance. In medieval alchemy it was used as a technical term for the process during which one metal showers another with effusions. The word's connection with the workings of the magic lantern and its shaping of our ideas of mental activity came later. By the end of the sixteenth century, "projection" named the mental process of constructing projects, planning ahead, and scheming futures (Schumer and Zubin 833). The magic lantern became a widespread influence behind western notions of intangible mental processes.

The delusory quality of widespread, mechanically projected images was accepted into eighteenth-century English as an analogy for a deceptive character type. In his picaresque romance *Ferdinand, Count Fathom* (1753), Tobias Smollett describes "the travelling savoyards who stroll about Europe, amusing ignorant people with the effects of the magic lanthorne" (174). Smollett also frequently uses the word "projector" to describe a posturing fraud engaged in a deceptive scheme of seduction or enrichment (Smollett 196, 212, 214, 272, 384, 398). Mechanical or human, a "projector" puts forth a fraudulent image.

By the nineteenth century, visions, demons, and gods were increasingly being explained away as optical delusions that are, as Freud would later put it, "nothing but projections." But well before the scholarly theories of psychology and philosophy, the idea that certain notions are nothing but projections appeared in nineteenth-century popular culture, literature, and scientific theory.

By 1808, an English political broadside criticizing civil liberties for Catholics depicted a magic lantern projecting a drawing of the pope and displayed the lines:

"What is this spectre of affright
With which they would delude our sight?
A shadow thrown upon the wall
A magic-lanthorn-shew! That's all!" (qtd. in Barnes, *Optical*)

The magic enchantment of the lantern is that it reveals the inner, hidden truth behind delusory, projected images. What a nice idea for subject/object thinkers to borrow from a machine!

Spectators throughout the nineteenth century could go home in awe from a performance by Robertson, Pepper, and many others, only to read later that it was all "nothing but projection." Technical journals such as the *Optical Magic Lantern Journal* explained how magic lanterns worked. Awareness of the point spread: *Supernatural forms are nothing but delusions created by the natural magic of projection from a machine or the subjective psyche.*

Usually implicit and hidden, the influence of the magic lantern on the notion of the subjective, contained psyche was at times made explicit. In the eighteenth century, a French satire pictured a monkey lecturing to animals, having forgotten to ignite his magic lantern's light (Florian). In a similar English satire of dull, pompous lantern talks in 1865, a lecturer also forgets to light the lamp inside his magic lantern. Nevertheless, he proceeds to bore half the audience of animals to sleep, blindly giving a slide lecture without the slides:

The spectators looked into a deep darkness,
Opened their eyes wide, yet were unable to see anything.
[He had forgotten at the beginning
To light the little lamp inside the magic lantern.]
He spoke eloquently, neglecting not one point.
But he had forgotten only one thing!
To light the light inside his lantern-head. (Florian)

The psyche is explicitly compared to a magic lantern; the lantern-head is like a projecting machine. This image appeared again in the cover illustration for the September 1980 issue of *Psychology Today,* in which psyche is represented as a movie projector contained in the skull.

The magic lantern's influence is not merely a historical curiosity; it had aided in the larger Enlightenment project of disenchanting the world, sweeping it clean for mathematical calculation and invention to serve human power.

In this world picture, the subject projects not only illusions but the framework of "nature" itself onto the world of objects. In research, a sphere for analysis is opened up not only by a methodology but by a prior fixed realm. Heidegger argues that a fixed ground plan must be projected forth to establish a binding system for research:

> This is accomplished through the projection within some realm of what is—in nature, for example, of a fixed ground plan of natural events. The projection sketches out in advance the manner in which the knowing procedure must bind itself and adhere to the sphere opened up. This binding adherence is the rigor of research. (*Question* 118)

This ground plan for research *projects* what is collectively assumed in advance, such as the system of mathematics and the self-contained plan of nature, from the model of the atomic structure of physics up through the various branches of knowledge: "Every event must be seen so as to be fitted into this ground plan of nature. . . . This projected plan of nature finds its guarantee in the fact that physical research, in every one of its questioning steps, is bound in advance to adhere to it" (119). This projection of a predetermined ground plan of nature reinforces the mystique of the world of subjects dominating the world of objects that seem to be "out there."

Projection is an important mental tool for modernity, setting up a realm of objective knowledge and then withdrawing subjective feelings that do not fit the ground plan of that knowledge. But there is a circular logic to projection's system, a self-determining method of knowing that reinforces what is already known by excluding what does not suit the plan. Thus we end up with unexpected side-effects such as an ecology crisis, because it was excluded from the projected plan of what was considered known for certain about nature. The subject seems like a tight container, but when we descend into the depths of the soul and the world, we find the notion of subjectivity to be bottomless. All the passions and images that the theory of projection attempts to squeeze into the objective psyche flow from endless, mysterious, bottomless depths of existence itself, more primordial than the subject/object divide and its mechanical models.

4 Streamlined, Sublime Speed

Speed has become the new goddess of the modern world.
—Jean Brun, *Les Masques du Désir*

Lust for the Rush

The world has picked up speed. As ancient Greek tribes trudged slowly across rugged landscapes, Icarus and Daedalus flew in their mythic imagination. Indigenous Inca and Iroquois runners prided themselves on their speed as messengers between communities. The Romans raced their horse-drawn chariots around hippodromes. Arabic sailors sailed before the winds, trading from Cairo to Indonesia. The nineteenth-century Pony Express raced to deliver mail across the North American continent at heart-pounding speeds, only to be replaced by the rapid-fire telegraph. The steam train stormed along like an iron horse. The bicycle rolled in a new horseless power for daring downhill racers risking "bone-smashers." The automobile captured the imagination of speedway fans. Huge ocean liners raced for the fastest time crossing the Atlantic. Airplanes excited frenzied crowds as the Wright brothers, Charles Lindbergh, and Amelia Earhart turned the ancient dream of flight into working machines. Carnivals still tease with roller coaster rides that swiftly spin screaming riders upside down. Rockets blast upward at thousands of miles an hour, sending robot spaceships millions of miles into space.

Modernity embraces a mad lust for the thrill, the *frisson*, the rush. It is not necessary, practical, or safe. It is expensive and polluting. But above all, it is sublime. Speed has become a form of ecstasy, a self-indulgent, high-pressured, reckless, and dysfunctional focus of technology. In *Faster: The Acceleration of Just About Everything*, James Gleick catalogs numerous ways that our culture has speeded up life. "Hurry sickness" has brought us industrial efficiency studies with stopwatches, pressuring workers to work faster; expectations of speedy travel resulting in traffic gridlock, road and air rage; "instant coffee, instant intimacy, instant replay, and instant gratification" (13). We suffer from increases in sleep disorders, short attention spans attuned to fast-cut television editing, increased expectations for rapid communication with e-mail, cell phones, and a faster-faster-faster-faster pace of life that crashes in a new medical disorder called "stress."

The desire for the rush is rooted deeply in human consciousness and is more exciting than the merely rational effort to improve transportation and communication. That sensible goal is constantly overshadowed by an archetypal desire for speed, which has long been germinating in the human soul, expressed in ancient horse races. Now the desire for speed is bursting forth in sublime, blazing machines streaking across the landscape and the sky—and clogging our "expressways" with traffic jams, sickening air pollution, hundreds of daily deaths in crashes, and drivers exploding in road rage. The technologies of speed are far from simply neutral, utilitarian machines. They are many-layered embodiments of the lust for the rush, the passion for sublime speed, the stressful chasing after ecstasy in speed. The astounding is transformed into the expected by the ego greedy for power.

This lust is to some extent a provision of ersatz satisfaction of deep unmet needs repressed by industrial consciousness. Speed is often promoted as an escape to relaxing vacations, to compensate people for the dehumanizing effects of industrial mass production. But there are more layers to this phenomenon. The enchantment with speed exceeds rational limits and becomes an obsession worth great expenditures and many deaths. It has become an idolized absolute, a deep craving for the power and thrill of conquering space and time. It fosters the illusion of exceeding the limits of mortality and approaching the realm of divinity like the Greek god Hermes zooming across the sky. The lust for the rush covers the archetypal range of desires, from crude to transcendent.

Fast machines now free the ego to feel like a winner, first in the line of cars. Daily excesses of the enchantment arise in speeding tickets, pressure to use cell phones while driving, and drivers enraged at barriers to haste. Racing-car drivers killed in action are elevated to folk-hero stature, as is demonstrated in the roaring death of the popular driver Dale Earnhardt. He was declared "The Intimidator" and lavished with praise by fans as if he were a bullfighter or boxer. He was killed on the race track in 2001, going about two hundred miles an hour. Leigh Montville's biography of Earnhardt is entitled *At the Altar of Speed.* Caught in this obsession, we are kneeling on the speed track, worshiping a new idol, because speed is intoxicating and ecstatic. It elevates its fanatics to participation in the excitement of feeling transcendent or sublimely godlike.

Kneeling at the Altar of Speed

One of the great visionaries of speed was the Italian writer F. T. Marinetti (1876–1944). In 1909, when automobiles and airplanes were rapidly developing the pace of travel beyond the horse and buggy and train, he initiated the futurist movement with his flamboyant essay "The Founding Manifesto of Futurism." Marinetti praised technology as the boundless wave of the future: not only would it sweep aside old traditions and explode in violent powers, but technology, he rightly saw, would saturate minds with a dazzling enchantment. Marinetti believed that futurism would cast aside ancient myths and promote a new, rational worldview: "Let's go! Mythology and the Mystic Ideal are defeated at last" (20). He happily trashes the old world's traditions: "We will destroy the museums, libraries, academies of every kind, will fight moralism, feminism, every opportunistic or utilitarian cowardice" (22). In his aggressive, macho proclamations, he obviously misread the future of feminism.

Marinetti's feverish pride was a preview of coming technological enchantments. He praised the love of danger, energetic fearlessness, audacious revolt, and the pugnacious "punch and slap" (21). Marinetti naïvely extended his pushy stance to celebrate war, which was another insight into technological culture's passions: "We will glorify war—the world's only hygiene—militarism, patriotism, the destructive gesture of freedom-bringers, beautiful ideas worth dying for, and scorn for woman" (22). World War I did not dampen his fever, and his macho militarism easily extended to welcome dictatorial leaders, for in 1929 he wrote admiringly of Mussolini ("Portrait" 158). Marinetti also foreshadowed our continuing aesthetic obsession with speed in the form of compulsively stylized cars. He glorified speed as a new beauty above Greek art, elevating the man at the wheel hurling his machine across the planet, adorned with pipes like serpents exploding threats ("Founding" 21).

Marinetti's passion for speed identified a widespread technological enchantment that gleefully swept aside serious ethical concerns. He promoted speed as a new religion and morality. In his 1916 essay "The New Religion-Morality of Speed," in the midst of the First World War, Marinetti argued that technology would replace religion. He rejected Christianity's morality and spirituality as irrelevant, now surpassed by speeding cars (94).

This visionary futurist condemned slowness with a scornful, violent ferver that would stimulate the creation of faster vehicles in the future. He urged the torture and persecution of any who "sin" against speed. He saw slowness as criminal, passive, pacifistic, and stagnant resistance to new speed ("Religion-Morality" 95). Marinetti's vision rises to the challenge and endows speed with a new divinity:

If prayer means communication with the divinity, running at high speed is a prayer. Holiness of wheels and rails. One must kneel on the tracks to pray to the divine velocity. . . . One must snatch from the stars the secret of their stupefying incomprehensible speed. . . . The intoxication of great speeds in cars is nothing but the joy of feeling oneself fused with the only *divinity*. Sportsmen are the first catechumens of this religion. Forthcoming destruction of houses and cities, to make way for great meeting places for cars and planes. (96)

Marinetti foresaw the power of our enchanted intoxication with speed. He foresaw the clearing of old sections of urban life to make way for modern expressways and airports. But, more importantly, he envisioned the modern worshipful kneeling on train tracks, race tracks, and runways as sacred precincts of the holy altar of speed. He envisioned the future's magnificent yearning to snatch the secrets of speed from the stars, which has led to the incredible velocities of space vehicles blasting above the earth at thousands of miles per hour then hurtling faster out into the solar system, fulfilling an ageless dream of the mastery of stellar travel. Imagine his pleasure if he had lived to watch landings on the moon or Mars. At the same time, we should not be surprised that he fused his passion for speed with crude male chauvinism and explosive fascism, because they express the crude, aggressive, deep power drive behind the collective lust for speed.

One irony of Marinetti's foresight is that he believed futurism would cast aside all myths and mysticism in the rush to technological dominion. He did not perceive what most technological enthusiasts exclude from their intellectual radar: the sweet technological worldview is in bed with charming myth and mysticism. Marinetti's sweeping enthusiasm for new machines, his intoxicating fascination with fascist violence, his lust for dominating macho power, his fantasy of speed as a new religion—these *are* the drivers in the seat of his race car, not reason. The speed machines that he envisioned did not by any means eliminate myths and mysticism. On the contrary, Marinetti represented a whole new era of futurists who drew upon old myths like Hermes to create a new mysticism of machinery. As if living in a beloved fairy tale of giants leaping across the landscape, hungry for godlike power, Marinetti and his followers were enchanted by technology. They thought in a trance, dazzled with technology's future. Marinetti's futurist descendants, even if shorn of macho fascism, still exalt speed so much that the auto and aerospace industries gleefully build machines that give drivers and passengers that addictive sensation of power and domination. The lust for being first on the highway is only a mild aspect of the intoxication with speed. That rush has driven many huge budgets and many complex engineering systems for fast ships, airplanes, hot rods, race cars, speedboats, and spaceships, and does not want to slow down.

Industrial society's rage for speed is only one example of the enchantment with technology that belies the conventional surface belief that our machines are purely the product of objective decisions and rational engineering choices, shorn of imagination, myth, mysticism, and enchantment. When we dig below the surface of rationalist consciousness, we release a burst of colorful and gripping passions that we normally suppress, but which unconsciously have their hands on the controls of our machines. These are told in the myths—not errors, but the meaningful, passionate dramas—and the mystifying dreams, like science fiction, that have hardly been swept aside by rationalism. On the contrary, the enchanting dreams sleep with the logical premises, the first principles of our grand technological enterprise. Reason loves them and gains ecstatic pleasure from these compelling charms. Drivers want to feel like rogues, outlaws, and unrepentant sinners in their beloved rumbling, rolling beasts. But these dreams can erupt into massively destructive violence, killing hundreds of people daily, smashing to painful smithereens our rationalist, futurist dreams of masterful control of ecstatic speed. Perhaps the greatest illusion of rationalism is that it can control the passions driving its speeding technologies.

Untamed Wheels

The automobile is perhaps one of the most unstoppable technologies invented, because it bestows an unparalleled individual freedom, mobility, and speed. It has become a powerful symbol of identity that industrial society's citizens are loathe to relinquish. In the United States we invest huge budgets in vehicles, highways, parking facilities, insurance, and fuel, and we tolerate thousands of annual traffic deaths and massive, dangerous pollution, just for this glorious freedom of power and speed. We fantasize about and fall in love with our cars, and we get thrills from their speed, even though in reality we sit in increasingly long hours of traffic jams, breathing toxic fumes. "The romance of the road pervades our fantasies," writes Jane Holtz Kay, author of *Asphalt Nation: How the Automobile Took Over America and How We Can Take It Back* (18). "It's not a car. It's an aphrodisiac," whispered a seductive 1992 Infiniti ad, echoing ancient parallel traditions: It's not just a rhino horn, or a bull's penis, it's an aphrodisiac. The make-out wagon and backseat bedroom repeatedly excites generations of new teenage drivers just waiting for the privacy of an enchanted evening. "Sex, near-death experiences—no wonder we are so emotionally attached to our cars," says the performance artist Laurel Guy. We blithely deny the damaging effects of our vehicle enchantment because we are so caught up in the pleasure, convenience, and thrills, even though the deaths may leave mental scars. Exploiting this obsession, U.S. corporations invest about $11 billion a year to promote cars. The automobile is not merely a useful tool for transportation; it is an expression of massive collective fantasies and desires. According to James Healey,

> What distinguishes the car business is passion. . . . [P]eople who buy cars and trucks remain emotional and involved. . . . Pop culture has idolized autos in song, from *In My Merry Oldsmobile* to *Little Red Corvette* and *Mustang Sally*. If the Beatles were bopping today, it's hard to imagine they'd replace the lyric "Baby, you can drive my car" with "Baby, you can move my mouse."

Just west of Salt Lake City are the remains of a vast evaporated salt lake that is large, level, and hard enough to support hot-rod car racers. It's called Bonneville. This is the Mount Olympus of hot rodding, where the "need for speed" is still the driving force (Steward 75). Here, the reigning speed king rules:

> Al Teague blasts along the Bonneville Salt Flats at half the speed of sound—in a car he built in his garage. The 58–year old racer has been on the salt since 1965 and currently owns the world land speed record for wheel-driven vehicles at almost 410 miles an hour. We'll do the math for you. That's better than two football fields every second. (75)

Speed makes longer trips possible, so the car is often promoted in advertising as a getaway vehicle, rushing in shiny isolation through a beautiful rural landscape, deceptively far from the urban traffic jams where it actually spends most of its time. Ironically, it is imagined as the machine that takes us away from the machine world. But the trouble is, millions of other machine drivers have the same idea, rushing in and out of cities so they can live in peace.

Rush Jams

In the United States, rush-hour congestion and cost in the eighty-five largest urban areas ballooned in twenty-two years. The average annual delay time per peak rush-hour traveler grew from sixteen hours in 1982 to forty-six hours in 2004. The annual financial cost of traffic congestion leaped from $14 billion in 1982 to over $63 billion in 2004. The wasted fuel from engines idling in traffic totaled 5.6 billion gallons in 2004. That much gas would fill 144 super-tanker ships, or 570,000 tanker trucks, end-to-end from New York to Las Vegas and back. That amounted to seventy-four gallons per traveler a year. The average cost of wasted fuel per traveler in all eighty-five urban areas was $829 in 2002. These figures do not even include the cost of building more roads and of pollution impacts, which is billions of dollars more for the nation.

How would increased public-transit ridership decrease congestion? New York already has the largest public-transit-rider percentages, and it still has significant traffic congestion. Other very large urban areas, such as Los Angeles, would have to increase public transit by over 20 percent, and small-to-medium areas, with populations under a million, would have to increase bus and train usage by about 80 percent (Texas). Traffic jams are like a plague on wheels. Transportation designers learned early that the more roads you build without also improving mass transit, the more traffic fills them (Kay 207). In 2001, the Texas Transportation Institute published a study concluding that "rush hour" is a laughable misnomer: then commuting time averaged three hours each way, including delays. Drivers are so enchanted with the apparent freedom of our automobiles that we are blind to the collective congestion, time wasted, fuel costs, and pollution that we are causing. And the number of cars in line is increasing. It's not rush hour, it's jam hour.

In Atlanta, despite billions spent on newer, wider highways, increased production of cars clogs them even more. Between 1992 and 1999, the time that the average driver in Atlanta was stuck in traffic delays increased from twenty-five to fifty-three hours a year. This cost the city $2.6 billion annually, or an average of fourteen dollars a person per workday, ignoring environmental costs (Krugman). If drivers were charged fourteen dollars a day to commute, plus environmental costs, they might reconsider, but the political will to make users pay their way and improve public transit is choked by our car obsession, which is fed by advertisements for new cars. This is an everyday, obvious, denied enchantment, a megamistake reaching to the sky in global warming.

Social Rank by Car

The cost, speed, and power of cars make them images of social rank, and auto marketers are quick to exploit this, purposely making some cars luxurious and others inexpensive. Consumers pay more for cars to enhance their place in the imagined social pecking order—the limosine, Mercedes, Rolls Royce, or Cadillac. Cadillac was the ranking U.S. status car in the 1950s, when long, heavy, powerful, luxurious, chrome-drenched cruisers advertised their owner's wealth in a society where badges of honor could be bought. The Oldsmobile was placed just under the Cadillac, and when Olds announced that it was closing down the brand in 2001, an article in the *New York Times* analyzed its enchantment: "A man might leer at a Lincoln or covet a Cadillac, people used to say, but he married an Oldsmobile" (Bragg). One elderly fan recalled his step up from Chevy to Olds, which made him feel "'on top of the world'" (qtd. in Bragg).

The speed of technological change allows car buyers to play with their automotive past. Nostalgia drives a retro car design niche, with the resurrected Volkswagen Beetle of the 1990s and the Chrysler PT Cruiser striving to look like a 1930s sedan carrying gangsters with tommy guns. Youthful drivers choose the Jeep-like vehicles, with their military background image. Like the ubiquitous student backpack, the Jeep says: "I would rather be camping (or recklessly tearing up backwoods trails)."

Sex on Wheels

When the Pennsylvania Railroad had the industrial designer Raymond Loewy streamline their locomotive, it was called by one critic "a gigantic decorated phallus" (Meikle 184). New fast train design is more subtle, but in auto marketing sexiness is never out of date; it only changes form. During the 1920s, the overthrow of stifling Victorian morality by those who might not have yet read Freud but knew all about new condoms, was already taking place in the back seats of the new Model Ts and other cars. Strict oversight of dating couples was now a thing of the past. Cars gained a sexy image as they became make-out couches on lovers' lanes. Joyriding couples threw old morals out the window, driving as fast as they could away from the old-fashioned chaperones to park their cars and secretly enjoy the unmentionable—necking and petting. In the 1920s, California citizens complained about lovers' lanes full of numerous parked cars housing "orgies" (Flink 159–60). The rolling bedroom still rolls along, leaving romantic memory traces in many a lover's mind. I recently saw a shiny red hot rod, its sides painted with images resembling sperm. "Sports cars are regularly described as voluptuous or testosterone-charged," says Phil Patton ("Proud").

The car's sexuality became explicit in design when the 1953 Cadillac was tagged the "Dagmar" because it had front bumper guards obviously shaped like swollen breasts. Dagmar was a blonde TV actress known for her ample chest. A 1950s ad shamelessly encouraged a subliminal phallic fantasy: "Make a date with a Rocket 88" Oldsmobile. In the 1990s, the new sports utility vehicles developed a more masculine look with bulging fenders and hoods shaped like weightlifters' biceps. Jerry Hirshberg, a General Motors designer in the 1960s, admitted,

> Back then, machismo filled the air at General Motors. It was certainly sexual, but it was the sexual fantasies of men. When we lapsed, we were doing design

pornography. . . . Today, we have muscle cars. . . . Nissan's first sport utility vehicle, the Pathfinder, was nicknamed the "hardbody," and designers described the "triceps" around each wheel. (Meredith 3)

Jerry Palmer, a designer for G.M., said that, male or female, "'people still like vehicles that are sexy,'" and today's popular models "'are masculine but they still have that sex appeal.'" Edward Golden, a Ford designer, said, "'A lot of women are buying trucks and sport utility vehicles, and I believe that they don't want to be seen as the weaker sex. . . . They want to be seen as equals'" (qtd. in Meredith 3).

But how could a machine—an object—be sexy? To talk this way you have to first think that a vehicle is not simply a practical, utilitarian technology for transportation, an object in a purely mechanical world. If that were true, any ugly machine that moves fast would do. Obviously imagination, symbolism, and passionate enchantments coming from the deep well of soul-in-the-world drag us into participating with cars as images, not only as clunky machines. They are art as much as engineering, as auto designers and marketers know well. Emotions affect thought. Anger, fear, self-confidence, and memories drive the size, expressiveness, and cost of weapons, clothing, homes, and other technologies. Marketers seek to establish "emotional branding"—a relationship with products that promotes consumption, from drinks to cars (Norman 60). A car is a moving sculpture, a collective image of unconscious desires, from crude sex and violence to social rank, refined elegance, and sophistication. Nor are these merely subjective feelings projected onto the objective machines. These desires give strong direction to the conception and design of cars: These consumer products are built and marketed to embody sexiness and pride. Cars transgress the subject/object dichotomy; they are hybrid phenomena, soul-in-the-world, passions on wheels.

Road Warriors

In the United States, there are about forty-three thousand fatalities each year from car crashes. Although police have enforced safety-belt usage more, the number of crashes has hardly decreased; 20 percent of fatalities are bicyclists and pedestrians (Kay 184). Car crashes, especially deadly ones leaving family and friends as corpses alongside the road, impact many sore hearts long after the metal stops screeching.

The road warriors' arsenal has been amplified by the sports utility vehicle, a bulked-up station wagon disguised as a truck that not only evades pollution regulations but appeals to drivers who want a big car to protect them in case of an accident, not a wimpy ecological car. What they ignore, or quietly love, is that their big cars are offensive road weapons as well. It is a case of *attitude*. Ellen Goodman says:

> The big cry from SUVers was "self-defense." Does that sound like Pentagonese or what? The most irate Suburban and Durango driving men sounded as if I wanted to take away their right to an assault weapon. . . . "If you choose a small death trap to drive, fine. If I want to pay the extra fuel and cost of an SUV, then it's my choice, not yours!" (Goodman)

Visit your car dealer's lot and you can feel the lust: "I want my SUV, the bigger the better—and never mind the cost of gas!" Despite vociferous opposition from environmentalists, Detroit is rolling out these monsters—3.5-ton, V10–powered, nine-passenger, ten-mile-to-the-gallon beasts. One such SUV, the Ford Excursion, was christened the "Exxon Valdez of Vehicles" by the Sierra Club. A major cash cow for the auto companies, the SUV has been a favorite of many drivers enchanted with its size and power and willing to ignore its problems: they devour fuel, crush smaller vehicles, roll over easily, and belch enough poison gasses to kill a small mammal. But U.S. drivers fell in love with this Big Dumb Car, intensifying the country's waste of resources for the pleasure of a few, while most of the world drives far smaller cars. Yet there are even worse polluters on the roads. Commercial diesel trucks and aging passenger vehicles are gross polluters. They continue to foul the air, generating 30 percent of the toxic smog, conveniently excluded from emissions regulations (Reed 9–11).

We need these hideous beasts, according to Hayes Reed, because they inspire us. They are not rational—they are overweight, oversized, go faster than the legal limit, waste resources, and pollute badly—but we need the fantasies they inspire, their enchantments. We need *inspiration*. Wait. Spiritual motivation from *technology*? Well, of course. Just like weapons and skyscrapers, our technologies embody our desires, crude or refined, and we are embraced by the world we create with them. We make them so we can be proud of them, love them, and let them inspire us. Damn the waste, the pollution, the danger, and the cost. We need that inspiration! We play with our technologies in a trance, role-playing dreamy characters we wish to be. Reed says:

> We wear our vehicles like costumes, using them to project an image that usually has no basis in reality. In a nation of serene suburbs and cubicle jobs, driving an SUV says, "I am not boring. True, I'm going to work now, but later, when you're not looking, I will be whitewater rafting, mountain biking, and climbing things. I am sexy and *dangerous*." Americans love SUVs because we are a nation of poseurs. We love to buy things that will never really have any practical

application—things like cowboy hats, pit bulls, Corvettes, assault weapons, etc. That we will never use them for their intended purpose is beside the point. The point is that they fulfill the need to *pretend*. (Reed 11)

We wear our vehicles like costumes? Why? We need to pose, to pretend to be sexy, dangerous, and bold—these fantasies are buried in the souls of nice middle-class Americans who must be itching to play the Hulk, to dominate the road, because these features have been built into these bulky, hunky vehicles. Their desperate collective souls needing to play roles are built into this steel, glass, and plastic, fueled by lots of gasoline. Hey, who cares? It's satisfying.

General Motors builds a Hummer, a consumer's version of the army's Humvee that is "truly meant for Road Warriors," according to the *New York Times* (Bradsher, "GM"). Looking like a smaller, shinier, squared-up military luxury vehicle, the passenger truck combines ruggedness and luxury for a mere forty-five thousand dollars and offers fourteen miles per gallon in the city and sixteen on the highway ("Hummer" 8). It weighs as much as three small cars. "'I love the fact that the Hummer is a tank; it's like a tank with fashion, it's like having your own war toy. . . . I like something where I can look down into another car and give that knowing smile that "I'm bigger than you." It makes me feel powerful,'" said Cooper Schwartz, the co-captain of his high school football team (qtd. in Bradsher, "GM"). This enchantment with power flattens the facts. Producing over five times the smog-producing gases, SUVs kill about three thousand people a year needlessly, including its occupants, who risk a higher rate of paralysis than drivers of smaller cars. This tanklike vehicle is intended not for going off-road but for the conspicuous display of consumption and satisfaction of reptilian domination instincts. Owners of Hummers certainly are successful—at buying war toys.

Clothaire Rapaille is a trendy French pop psychologist of consumerism. An amateur Jungian, he sells his insights in workshops, books, and his newsletter, *Archetype Discoveries Worldwide*. The Hummer is a car from the reptilian part of the brain, according to Rapaille; it appeals to archaic instincts of strength and bad-guy aggression (Barrett, "Hummer"). The United States is an alpha-male culture, he says, big, strong, and number one in production, weapons, and money—the "hyper power in charge of the planet," the new Napoleonic France. At an auto show, Rapaille was struck by all the big engines—a ten-cylinder, five-hundred-horsepower motorcycle (a "crotch rocket," said one designer); a ten-cylinder Rolls Royce; and a sixteen-cylinder, one-thousand-horsepower Cadillac aimed at gas-guzzling rich consumers. Some drivers see these cars as assertions of patriotic confidence: driving such cars is a God-given American right, an assurance of U.S. technological superiority, a "power-junkie passion." Rapaille says that big engines may be a response to fearful crises. Threatened, "the reptilian brain says, 'I want power, I don't give a damn'" (Patton, "Proud," 1–4).

Cars are statistically more dangerous than guns. They are weapons regularly used in homicides and suicides. A suburban teen boy is more likely to be killed by a car than a city boy by a gun. Media images romanticize racing, machismo, and leave safety out of the picture. Add food, cell phones, alcohol, and drugs to the scene, and you have thousands of deaths annually (Kay 26, 107). Rational transportation is overwhelmed by aggressive drivers shooting anyone who gets in their way. A new breed of "driving psychologists" asks, "Could it be that people drive in their imagination more than on the real road?" (James and Nahl 255) They have identified numerous factors contributing to road rage, from loud music to the Jekyll-Hyde Syndrome, passive-aggressive road rage, the Rushing Maniac, the Aggressive Competitor, and the Scofflaw (James and Nahl chap. 4).

In the richest nation in the world, with the fastest cars on the road to whisk us home after work, why can't we be patient, considerate, and polite to other drivers? Contempt, says a New York City driver, Lynette Abel. Contempt, she says, is a false sense of self-importance and glory derived from diminishing others. Contempt seems to be bred by car-inflated egos. Ellen Reiss responds:

> "As car doors are opened and engines are turned on, millions of people, including well-behaved people, feel deeply, 'Now some power is in my hands. This world that confuses me and seems so often to thwart and stop me, I can literally ride over now—I can step on this accelerator and feel I'm finally unobstructed, having my way. As I go over this road fast, maybe cut in and out of traffic as I please, I feel finally I'm conquering the world, I've beaten it out.' . . . [O]n highways and city streets, drivers are cutting off other drivers and yelling things at someone which they would not think of saying if they saw that person in their office or even in a store." (Qtd. in Abel 8A)

When cars are raced down the four-lane expressway by enraged, contemptuous drivers, they are not simply objects "out there" on the road. They are carefully crafted cultural artifacts, technologies made with engineering skill to be driven by explosive passions such as aggression, in a cultural context that loves it all. Speed, anger, and the lust for power and domination are *built into* these vehicles; these desires are the motives for shaping certain types of car, such as the SUV. These car-weapons are steel and plastic soul-in-the-world, a union of mind-body phenomena—not neutral objects but stars in the enchanted drama of speed.

Car Welfare

Enchantment with the automobile is so entrenched that the U.S. government subsidizes it with major infrastructure financing. Local road projects are often funded up to 90 percent by federal grants, while support for mass transit must come largely from the smaller budgets of local sources. Cars and trucks receive *seven times* more government support than mass transit: "Since World War II auto-centric policies have eroded the nations's public transportation policies, now receiving one government dollar for every seven handed to the car" (Kay 44). Urban public space is dominated by automobile infrastructure: "From 30 to 50 percent of urban America is given over to the car, two-thirds in Los Angeles" (64). Billions of federal dollars spent on expressways equals car welfare, but drivers dazzled by convenience are blind to the collective costs. There is no such thing as a "free"-way, but the name reinforces the illusion of auto freedom and the denial of costs.

The U.S. taxpayer subsidy for automobiles and trucks exceeds education and health expenses. There lie our priorities. We drive in expensive vehicles, use products carried by expensive trucks, all on expensive highways, but we are dazzled by our illusory fantasies of freedom. Imagine if this information were posted on billboards: Car drivers require an average of six thousand dollars of their individual costs a year and another three to five thousand in taxes for the public costs of parking, police, environmental damage, uncompensated accidents, and land devoured by sprawl. Add about ten thousand dollars a year to your car costs, folks (Kay 120). The cost of our car infatuation extends to international relations, especially regarding oil, which consumes 6 percent of the U.S. trade deficit. Auto parts make up 66 percent of our trade deficit with Japan. The military spent about $50 billion a year to defend Middle East oil reserves before the invasion of Iraq in 2003. That is about two hundred dollars per citizen in taxes annually (and rapidly rising) for weapons to defend a gasoline source.

An obvious solution is mass transit, which is more economical and less polluting. Buses, trains, ferries, and monorails could easily be built instead of highways, at less overall cost than the large numbers of cars driven by single drivers. "A billion dollars invested in mass transit produces seven thousand more U.S. jobs than does the same amount spent on road construction," and it promotes more long-run economic growth (Kay 129). Yet investments in New Jersey's highways alone in 1990–94 exceeded Amtrak's train expenses nationwide. Cars are subsidized by taxes so that drivers pay less than half their costs, but train passengers pay about 80 percent of their costs (Kay 128–30,

278). This is a massive enchanted system of politically popular but costly automobile welfare that suppresses sensible public transport.

The national highway system is neither economical nor safe. It was never vital to national defense, as its builders claimed, and it has cost far more per person than mass transit would have (Flink 214). Auto companies intentionally bought and dismantled public transport systems (220). Interstate highways destroyed neighborhoods and city parks and have still not alleviated rush-hour congestion, due to increased numbers of vehicles. Auto manufacturers at first resisted safety devices such as seat belts, costing thousands of lives (217). Motor exhaust continues to be a major cause of heart and respiratory disease (222). The United States, with only 6 percent of the world's population, owns over half the world's motor vehicles (231–32).

Auto culture has built an expensive castle of transportation that embodies passionate desires creating a sentimental dependence, a seeming imperative, an addiction that drives Americans to buy over a million cars every month. Why are industrial cultures so addicted to cars that they refuse to see the rational need to reduce car and truck usage and improve mass transit? Our fascination with cars is a love affair, a passionate dependence on the freedom, symbolic power, sexiness, and social status that cars embody. All the transportation pleasures and comforts that would have been the exclusive privileges of the rich a hundred years ago are now fanatically guarded as the necessary framework of the middle-class life of wealthy industrial nations. Car welfare is a major political ingredient in our auto obsession. We must dig below the problems of "safety last" (O'Connell) to grasp the enormity of our obsession with vehicular speed and take responsibility for our enchantments. The subject/object dualism falls short of explaining our collective car enchantments by neglecting the deep holistic union of desire that becomes costly concrete, steel, and pollution.

Solitary Isolation

Automobile speed has also fostered social isolation. Some people enjoy the isolation of driving alone, finding the car, in a life of hectic speed, to be their only solitary time in what Kay calls "the stillness of this chamber of isolation" (26). Isolated behind glass in techno-cocoons, why bother to care about "external" problems? The car enables the creation of the suburbs by enabling the commuting that allows building on the fringes of the city. This leads to further automobile-driven isolation. The solitary patterns created by cars have even led to fearful gated and walled communities in fortressed America, with the nearby mall as another enclosed, car-surrounded commercial shopping temple. Luxury gated communities number over thirty thousand, mostly in the southwestern United States, with entry by car only, heightening fear and intolerance of the "other" (Kay 71).

In addition to this sociological view, the psychological and spiritual view of isolation is that cars promote egocentric pride. The Amish, who reject automobiles, believe that cars fling people apart, exaggerate the desire for speed, and inflate drivers' egos. The automobile, driven to work by a solitary driver, is an isolating machine that fosters the taken-for-granted metaphysical assumption of the secluded subjective psyche in a dead, soulless world of objects. Contrast driving with bicycling to work, camping, or a walk in the woods and the feeling of real participation in the wider world, grand nature, the spectacular universe that such simple activities can cultivate.

Ignore That Pollution

"'People in this city would rather drive than breathe,'" complained the environmental official of Mexico City, where the air is full of toxic concentrations of ozone, bus exhaust, industrial smoke, and particles of dried fecal matter (qtd. in Preston 3). Mexico's capital is one of the most toxic places on the planet to breathe. The dangerous habits of pollution in highly industrialized countries are spreading rapidly to developing countries, along with the dreamy enchantment with speedy vehicles that promotes denial of the crisis. In Mexico City the residents know the symptoms: "'We just breathe pollution, so we never get well,'" said Ivonne Vega, twenty-three years old, as she pressed a tissue to her nose (qtd. in Preston 3). Her sniffling eighteen-month-old toddler was just recovering from one of his almost-constant throat infections. But vehicle drivers and other polluters refuse to give up their addiction to their machines:

> Alejandro Encinas, the city's Environment Secretary, said he receives nothing but complaints from drivers who have been forced to clean up their engines or whose dirty cars have been halted on high-pollution days. (Preston 3)

When Mexico City's bus companies proposed larger buses for ecological reasons, the drivers resisted; in the name of jobs and tight traffic conditions, they preferred more small buses (Preston 3). What irrational mania would compel Mexicans to let driving take precedence over breathing? Only the same enchantment that compels drivers around the globe—the passion for the power, freedom, and satisfaction of other desires overwhelms our ecological judgment.

Auto manufacturers resist making smaller low-emission vehicles and continue making profitable large cars and polluting diesel trucks. Drivers demand the freedom to pollute the air we all breathe, buying more and more highly polluting cars and trucks. Thirty thousand people a year die in the United States from respiratory diseases stimulated by our precious cars and trucks (Kay 111). Can you imagine drivers anywhere saying, "Oh, we are causing massive unhealthy pollution? Well, of course, we will cooperate in a massive cutback of these clouds of poison"? Less-polluting hybrid and fuel-cell engines are on the way, but long after older cars have seriously polluted the environment. The hybrid car engine was such a success with reducing gasoline use that its marketers took it in a contrary direction. Next they promoted new designs of high-horsepower engines with hybrid electric motors, added not to save gas seriously, but to give a boost to the power-hungry drivers who want large engines and fast speed. It does save a bit of gas, given that goal, but the purpose is not ecological benefit but speed and profit. Marketers know that horsepower outsells gas mileage, so instead of encouraging ecological cars, they are changing hybrids from ninety-six-pound wimpy cars into muscle machines: "Forget about those puny gas sippers. The hybrid in your future is all about hot-rod horsepower" (Naughton 50). Mass transit is even slower to gain political support so as to reduce the number of cars on the road altogether.

Just a few figures indicate the use of resources and pollution side-effects that, in our obsession, we ignore. The environmental cost of manufacturing a new car is over 33 percent of that created during its driving life. The processes of extracting raw materials and manufacturing the car is the equivalent of driving thirty-five thousand miles. Although U.S. cars have doubled their fuel economy and cut their per-car emissions, drivers also doubled the number of miles driven since 1980, so these gains were wiped out. In one second U.S. cars and trucks travel sixty thousand miles and use up three thousand gallons of petroleum products. In the mid-1990s,

> Our fossil fuel vehicles were . . . exhaling two-thirds of [U.S.] carbon dioxide emissions [sixty thousand pounds], one-quarter of its chlorofluorocarbons (CFCs), more than 50 percent of its methane, and 40 percent of its nitrogen oxides, plus most of the carbon monoxide. . . . [Y]ou are driving a pollution machine. (Kay 80–83)

Are you jogging along a road for exercise but breathing all these gasses from passing vehicles? Over a third of the U.S. population is breathing air that the federal government considers unfit. The thirst for oil also creates serious water pollution. Of the many toxic sources of water pollution, oil spills contribute a large share. Over thirteen thousand oils spills, some small and some massive, occur every year in the United States. In 1979, the Ixtoc I oil well in the Gulf of Mexico blew out *140 million* U.S. gallons of oil; in 1978, the oil tanker Amoco Cadiz ran aground in the English Channel and spilled 69 *million* U.S. gallons; and in 1989, the Exxon Valdez tanker spilled only 11 *million* U.S. gallons of oil. Massive pollution in spills from trains, trucks, pipelines, gas-station tanks, and ships carrying fuel oil is very difficult to clean up. Massive numbers of fish, shellfish, birds, and coral reefs can be destroyed for years after an oil spill. Certain frequently traveled shipping lanes, such as the Persian Gulf and the Bosporus Channel, are dangerous for oil tankers due to shallow seas or rough water. And some ships even *intentionally* dump waste and oil into the sea! Oil and gasoline linger in water for a long time and kill massive quantities of organisms. Car owners and home mechanics dump over a hundred million gallons of used motor oil a year into the ground, sewers, or waterways. One gallon of raw oil can contaminate a *million* gallons of drinking water, essentially forever (Kay 95).

The production, transportation, and storage of fuel and oil for our cherished automobile and truck culture is a massive system that is polluting the air and leaking and spilling into our drinking water regularly, sometimes on a disastrous scale. Yet the entrenched passion for speed and convenient travel steamrolls over these problems, ignoring, denying, and fighting against regulation and cleanup at every level, from industry to consumer. What will happen when developing countries such as China and India succeed at emulating the industrial countries' high levels of automobile consumption and pollution? Environmentalists in industrial nations fight an uphill battle against international pollution regulations as long as they are outnumbered by the gas guzzlers. Our addiction to speed undermines global environmentalism. "The hypocrisy of America preaching from behind the wheel cripples efforts to curtail consumption worldwide" (Kay 98).

Streamlined Idol

Perhaps the most effusive expression of the auto's meaning for industrial culture is streamlining, which is the imitation of airplane design in cars with smooth curves to reduce air resistance and to make them look like they are going faster. Some streamlining is practical, while some is purely decorative, more expressive of sleek elegance than utility. Streamlining signifies far more than speed, however, for its variations have taken ridiculous, purely symbolic forms, such as the 1950s tail fin. Streamlining began as an imitation of airplane and rocket design, so drivers could fantasize at some level that they were *flying* in the "cockpit" (Freud would love that image) of the speed machine par excellence, the new jets and rockets of the 1950s. The farther that car styling got away from practical function and sensible design, the more the car became an icon of all the passionate desires embodied in that machine. The automobile became a beloved image of the enchantments of the machine world, a self-moving *idol* worshiped with costly tithes. Auto idol worship began in the 1920s. It was permitted to rip up cities for larger roads and parking lots. Ford's "Tin Lizzie" was worshiped as a work of a technological savior bringing miraculous nonhorse travel to the masses (Kay 173). Race tracks, notably the Indianapolis Speedway, began the ritual of tinkerers refining engines for more speed. Daniel Boorstin calls this the worship of the "motor goddess":

> "If we are an automobile-riding, we are also an automobile-ridden people. Despite the daily offerings that Americans insert in parking meters, and the grand new parking temples rising in the centers of our cities, we seem unable to appease the motor goddess." (qtd. in Kay 269)

Much as a cathedral or a mosque embodies the aesthetic style of a religion, streamlining is the aesthetic style of the goddess of speed. Cars slowly grew out of their imitation of horse-pulled buggies but did not know how to look until the industrial design movement emerged in the 1930s, led by the graphic designer Raymond Loewy. Industrial design is a branch of modern art, a blend of art and engineering, making machines fit human needs and express their function. Loewy led the pack with endless striking expressions of speed wrapped around machines, some of which never moved. His designs display an elegant, streamlined simplicity, as if racing or flying, because he loved speed. "'I was a speed demon ever since I can remember,'" he admits (qtd. in Anderson 98).

Loewy streamlined the locomotive in the 1930s. The classic Streamliner S-1 engine, designed for the Pennsylvania Railroad, looked every bit like a Buck Rogers rocket on wheels. An advertisement called it "the shape of speed." His popular 1953 Studebaker design was low and sleek, as if the car were racing as fast as possible; his still-influential Greyhound Bus design was so smooth that the bus did not seem as bulky as it was. When he was seventy-five, Loewy obtained a racing driver's license. "'I once accused him of making streamlined pencil sharpeners,'" the architect Philip Johnson recalled. "'In our fundamentalist view, there was no point in streamlining something that didn't move. Now we appreciate streamlining as a valid movement in American design. And he started it'" (qtd. in Anderson 102). Loewy replied, "'Streamlining is reduction to essentials'" (qtd. in Anderson 102). Streamlining means that it is important to conceal clunky mechanical elements, such as the inner framework and engine of a car, in order to express the delight with speed with sleek wraparound steel covering. It also means evoking speed in autos by copying fantasy machines, such as rockets and airplanes. Some streamlining designers applied the look to radios, telephones, electric razors, and other machines, whether they actually involved speed or not.

Loewy later helped design the interior of space shuttles, insisting on considering the astronauts' human needs, such as sleeping on a flat surface instead of floating, and a window for them to see earth below, which they greatly appreciated. Loewy's revolution in industrial design was based on rational principles such as simplicity and function ("form follows function," industrial designers chant) and the soaring creative imagination of an artist (Anderson 103). Streamlining had a wide design impact beyond Loewy because it sprouted in an industrial culture that is fascinated, enchanted, and obsessed with the goddess speed.

Streamlining was so popular that it was easy to take it to absurd extremes. In Detroit, car "stylists" developed an extreme of streamlining that was uselessly decorative but a howling commercial success for a few years in the 1950s: the tail fin. They were led by Harley Earl, a Hollywood car customizer who was recruited by General Motors to stylize cars. "His trademark was to make cars look like they were moving fast—even when they were standing still. He replaced the boxy lines of the 1920s with sleek curves and angles" (Mingo 22). He says that he was inspired to put tail fins on cars when he saw a new airplane, the P-38, which had double rudders that soon appeared on the 1947 Cadillac in modest size. When their popularity grew, their size also grew, so that by 1959 the Cadillac's fins were large, pointed, and evoked a rocket image. In 1953 Earl saw a photo of a new F-4D Skyray with a delta wing, and he again applied airplane streamlining to cars. The fins grew into higher and wider absurdities of pure fantasy in steel. These gas-guzzling hunks of flying steel lowered the gas mileage of U.S.-made cars to about 13.5 miles per gallon in 1973, and one critic described these tail-fin monsters as "Huge, vulgar, dripping with pot metal, and barely able to stagger down the highway" (Flink 286–87).

As a designer (not a "stylist"), Raymond Loewy ridiculed the new cars as "'jukeboxes on wheels,'" and the president of American Motors, George Romney, called them "'dinosaurs in the driveway'" (qtd. in Flink 283). Earl responded with his purely commercial motives: "'Listen, I'd put smokestacks right in the middle of the sons of bitches if I thought I could sell more cars'" (qtd. in Mingo 25). By 1959, as Jerry Flint put it, "as useful as rust, tail fins were an outrageous rage." The absurdity was even worse than ridiculous. In New York City a woman died when a Cadillac, rolling backward, hit her with its sharp fin. In Chicago a boy impaled himself on a sharp tail fin and died when he ran backwards to catch a baseball. But car tail fins died around 1959 for other reasons: "It wasn't the danger that put a stop to fins—it was the laughter. They were so outrageous that they became the butt of jokes" (Flint 3). Jean Baudrillard saw tail fins as a sublime triumphalism of the machine, pure signs not of real speed but of the sublime imagination of speed, of an imagined victory over space. They implied a miraculous automatic machine gift from above. They played at propelling the car into flight, like an airplane, a bird, or a rocket (Baudrillard 59).

People of all cultures idolize absurd images that stimulate excitement. The tail fins on cars that took streamlining to this mindless extreme expressed a manic enchantment with the growing speed of industrial culture, just as new jets and rockets, satellites and expressways in the 1950s seemed to embody the age-old dream of acceleration's thrills. Cars that imitated jets and rockets gave drivers the pure unconscious fantasy of being pilots in supersonic space, of participating in their industrial culture's rush toward the future, which did turn out to be speedy but also dangerous, expensive, polluted, and stressful.

Some industrial designers reacted against this. Victor Papanek, who published *Design for the Real World* in 1972, criticized the wealthy industrial nations' indulgence in "sexed up" killing machines with tail fins and wasteful luxuries promoted by too many designers and marketers. He foresaw the coming global problem of overpopulation, urban density, and neglect of the impoverished majority of the world. He and his students designed ecological vehicles, a nine-dollar television, a nine-cent radio powered by a candle, and a six-dollar food-cooling unit for developing countries (Papanek 92, 190, 270).

Speed, Style, Metaphysics

Our vehicle enchantment is a bottomless immersion in a primeval sea of instinct and passion—sex, aggression, greed, lust for power, and exciting speed. Fetishized into an icon of freedom, expansive American destiny, glamour, and drama, a car offers with no particular strength or intelligence needed by the driver, a "sublime enchantment" (Benesch). Baudrillard sees this through Freudian theories of eros and narcissism: speed offers the thrill of transcendent power in the rapid mastery of space. Speedy machines stimulate a narcissistic, self-indulgent feeling of mastery and dominance that encourages the lifting of taboos such as safety limits, releasing one from responsibility in a cloud of daring ecstasy. Dynamic speed infatuates with its phallic, seductive thrusts past ordinary space and time. This narcissism is intimately erotic, removing resistance and offering enchanting thrills (Baudrillard 68–69).

These erotic passions are not simply subjective, interior feelings limited to the human brain. Cars are not only objects "out there" in a mechanical world; they are part of the cultural system that we create. Many technologies are "sexed-up" embodiments of our invisible passions. Pollution and wasteful consumption are materializations of our lust for the rush. Though cars are useful machines, they are taken to absurd extremes. We need all the objective, rational analysis that we can muster to manage this crisis, but we must also dig deeper into soul-in-the-world to find the roots of our enchantments. Then we can seek the solutions at a new ecological level, deeper than the subject/object metaphysic of industrial culture.

Looking back at modernism's excitement about the rising speed of its machines, we need to gain the ability to see what modernist subject/object metaphysics could not see. Modernists relegated symbolic aspects of technology to merely subjective, and therefore insignificant, decoration and thought of engineering as a rational analysis of an objective world. They failed to see the collective, nonsubjective, and nonobjective enchantments that possessed them. Captivated by the excitement of speed, designers were concretizing collective fantasies in streamlined trains that looked like fantasy rockets. Car stylists were embodying the speed obsession of the age in auto styling, shaping rationally irrelevant tail fins. A tail fin on a car is not a neutral object projected onto a car design by a subjective fantasy; it is a collective, archetypal fantasy, right there in steel, more sculpture than machine, a trite effort to induce the trance of speed.

Speed is a sublime rush, whether zooming on the highway or twinkling from a satellite speeding above in the night sky. Streamlining expresses our immersion in the enchantment of speed's excitement. But our technologies of speed force us to broaden our classic subject/object metaphysics. There never was a human subjective fantasy metaphysically separated from the objective facts of cars, airplanes, and rockets. These machines are built by the driving passions and collective manias of industrial culture—the lust for the highway rush.

Industrial societies don't choose to relieve global poverty by making lots of appropriate technologies—-simple tractors or water pumps for struggling countries. Rather, we invest heavily in our own passions for speed. We are making soul-in-the-world in those streamlined machines. We are participating in the delirium of speed as it guides us to shape expressions of that intoxication and then worship it as a goddess. This is not subject/object metaphysics. Industrial culture's metaphysical foundations are not what they seem to be. Rational engineering and enthusiastic art emerge into consciousness from a deeper place where they are one: passionate participation in the profound depths of existence itself.

5 Titanic, Utopian Triumphalism

> The illusions, aspirations, and whims of mankind, even more than its physical needs, influence profoundly the beliefs and activities of scientists.
>
> —Rene Dubós, *Dreams of Reason: Science and Utopias*

Paradise Adjusted

Triumphalism—the assumption that modern technology has conquered most barriers and is an unstoppable, victorious, utopian historical force—is a major enchantment of our technology. Cultures understandably want expressions of optimistic hope for their goals, but what will be the fate of technology's titanic triumphalism? We need to dig down into the archetypal utopian imagination to see.

Early utopias populated world mythologies, such as the Garden of Eden, Hesiod's Golden Age, Plato's Republic, and heavens such as the Christian City of God. They demonstrate the global, archetypal pattern that still pervades a great deal of technological thinking. In each era, utopias express different themes, adjusted to reflect the enchantments of the time. Soon after Columbus inspired Europe with new dreams upon his return from the New World, Thomas More's classic 1516 book *Utopia* set the tone for European tales of a remote island with socialist and technological advances. Late nineteenth-century and early twentieth-century utopias promoted highly technological dreams.

The nightmarish side of utopian technological dreams erupted in the *Titanic* tragedy, a classic event of utopian optimism turned dystopian. By the second half of the twentieth century, following the devastations of its horrible wars and totalitarian atrocities, utopian literature seems to have died after two major antiutopias: George Orwell's *1984* and Aldous Huxley's *Brave New World*. Not so. Technological utopia was only being adjusted. Technological utopias—some in science-fiction literature and cinema, some in the sober predictions—continue to glorify promises of moral improvements, pleasure, and power, but now they are more often mixed with dystopian horrors. Utopia always has a shadow nearby, whether it be a place such as hell, a legend such as the Fall, the totalitarian controls proposed to assure utopian principles, a disaster such as the sinking of the *Titanic,* or a science-fiction blend of futuristic technology with fascist terror, as in *The Matrix.* These are major enchantments of technology exploring the light and dark potentials of the industrial worldview's present concerns in futuristic settings.

Dominion and Francis Bacon

Utopian imagination pictures societies of grandly improved comforts, justice, and more recently, technologies. But ironically, utopias always have a shadow. One of the strongest themes in technological utopianism is control of technology and domination of nature. The control of nature through science and technology was a welcome goal in the Renaissance. Francis Bacon's utopia stressed the need to experiment rather than indulge in otherworldly medieval speculations. He wrote during the inquisition of witches in the early seventeenth century, and he maintained a firmly patriarchal stance. For Bacon, knowledge is power, and he is often quoted in his most aggressive mood, saying things such as, "'I am come in very truth leading to you Nature with all her children to bind her to your service and make her your slave'" (qtd. in Merchant 170). This appears in his essay "The Masculine Birth of Time" and indicates the highly patriarchal enchantment that colors the culture of slavery and the oppression of femininity. He often describes matter in female images, such as "a common harlot" seeking to overthrow his quest for a new order and return to an older "Chaos" (Merchant 171). For Bacon, nature is feminine and needs mastery:

> "[S]he is driven out of her ordinary course by the perverseness, insolence, and forwardness of matter and violence of impediments, as in the case of monsters; of lastly, she is put in constraint, molded, and made as it were new by art and the hand of man; as in things artificial." (Qtd. in Merchant 170)

Nature is seen as chaotic and in need of male control. As Caroline Merchant observes:

> For Bacon, as for Harvey, sexual politics helped to structure the nature of the empirical method that would produce a new form of knowledge and a new ideology of objectivity seemingly devoid of cultural and political assumptions. (172)

Bacon's thinking is ambivalent, however—on the one side, he seeks a more aggressive, masculine power over nature, and on the other side, he expresses a decent concern for the common human good: opposition to slavery, private gain, or destruction. His empirical research method is specified in his *Novum Organum* (1620), and his utopian picture of the outcome of technological developments is portrayed in his *New Atlantis* (1627).

The *New Atlantis* is his fictional South Seas island utopia of good will and advanced technology, guided by Christian religion. Lost sailors stumble onto the island paradise of Bensalem. Christianity was brought to this lost civilization, the inhabitants say, by a great pillar of light on the sea, and then it developed advanced technologies. Good will is so finely cultivated that social order is easily maintained. In this modified aristocracy, distressed families are supported at public expense. Disobedience is rare, due to the "reverence and obedience they give to the order of nature" (19).

Sexual behavior is so admirable that there are "no dissolute houses, no courtesans, nor anything of that kind" (23). But there is no mention of including women on an equal plane with men, except in equal inheritances. This social utopia is amplified by a commitment to research and technological development in Salomon's House, a modern technological research university that is run by men, the "fellows or brethren" (17). The impressive technological accomplishments of Salomon's House cure diseases; prolong life with "the water of paradise" (28); build half-mile-high towers; breed new types of plants, insects, and animals; study animals by dissection; provide abundant food and heat; study math; produce imitation smells and tastes (28–32); make weapons including submarines; and achieve "some degrees of flying in the air" and "some perpetual motion" (32). They enjoy the imagination with "houses of deceits of the senses," where they display "false apparitions, impostures, and illusions," although they "do hate all impostures and lies" (33). All these riches and technologies are cultivated with new experiments, and they honor their inventors. Above all, technologies are guided into beneficial use.

Bacon's utopia combines the common post-Columbus and imperialist tale of a socially ideal paradise found on a distant land with his vastly enlarged vision of a technologically advanced culture. His contribution to technological development was not only the experimental method. He also promoted a strong urge to aggressively dominate and master nature. This is evident in this utopia's high towers, plant and animal cross-breeding, dissection, and weapons. He is blissfully blind to the destructive side of technology that would intensify in the twentieth century. Bacon optimistically imagines human nature to be so cooperative and respectful that only moderate forces (such as censure and secrecy) are needed to maintain social order. Most important, the technological wonders are guided by a refined spiritual commitment to turning inventions to "good and holy uses":

We have certain hymns and services, which we say daily, of laud and thanks to God for His marvelous works. And forms of prayers, imploring His aid and blessing for the illumination of our labors; and turning them into good and holy uses. (35)

The rationality of technical thinking was not regarded as sufficient to guide technological applications. "Numberless in short are the ways, and sometimes imperceptible, in which the affections colour and infect the understanding," Bacon wrote (*Novum* 474). Thus, explicit and strong spiritual and ethical guidance keeps the affections at a quality level.

In *Novum Organum*, Bacon developed his experimental method. Pertinent to our study of enchantment is his naming of the hindrances to understanding the causes and effects of natural laws. These hindrances, or false opinions and prejudices, he calls the "idols" (*idola*) of the tribe, the cave, the marketplace, and the theater. These idols must be set aside to free the mind to discover natural laws experimentally. These false beliefs are illusions, Bacon argued, for nothing exists in nature except material bodies acting according to fixed laws. However, Bacon was blind to his own idols, especially the patriarchal domination of women and the new surge of the quest for power and control over nature. He may have seen these as valuable at the time, but they continued as concealed motives behind the ideology of objectivity. And as the power of technology grew, these idols became more problematic, for they could do more damage. Bacon's rejection of idols did not exclude God, however:

There is a great difference between the Idols of the human mind and the Ideas of the divine. That is to say, between certain empty dogmas, and the true signatures and marks set upon the works of creation as they are found in nature. (*Novum* 466)

Bacon must have hoped that his idolatries, or enchantments, would be openly condemned by scientists in future centuries, but little did he imagine the outwardly declining public influence of the religion that he counted on to guide technological development on "good and holy" paths. Nor did he foresee the great irony that technological reason would never rid itself of these idols hundreds of years later.

Bacon knew well that technology could be corrupted by human desires. But he urged its development under the direction of right reason and God, for God assigned humanity the right to dominate nature:

Only let mankind regain their rights over nature, assigned to them by the gift of God, and obtain that power whose exercise will be governed by right reason and true religion. (*Novum* 539)

But by stressing dominion, Bacon helped undermine the ethical and re-ligious guidance of technology that he advocated. Judaism and Christianity, as Lynn White has shown, have long helped disenchant nature and have pro-moted the dominion of humanity over nature by relegating the divine to a remote transcendent realm. He wanted to retain the best of the religion by calling on faith in God to ethically regulate the development of technology. However, simultaneously, he encouraged inventors and scientists to adopt the attitude of domination and control of a feminine chaos, which became an enchantment integral to the aggressive development of technology. These contradictory directives are still competing, with dominion running well ahead of ethical restraint.

Control in Utopia

Traditional utopias inevitably include elements of social control that could easily become dystopian. Control became even more potent as technologies multiplied. When traditional religion lost its grip on the social regulation of technology, promoters of utopian visions became ambivalent. The heart of western religious faith is the ultimate affirmation of good over evil. But without God to guarantee this faith, utopian visions of technological societies had no assurance that human nature alone would guide technology along benevolent paths. So utopian, triumphal hopes had no necessary ultimate reason to over-come dystopian horrors. To this day, utopian hopes for progress are darkly shadowed by dystopian fears of technological horrors. For each *Star Wars,* where the light side of the Force barely overcomes the equally strong dark side, there is a horrifying *1984, Stepford Wives, Alien, Matrix,* or *I, Robot.* For every technological monument such as New York's Empire State Building, there are millions of poor, resentful, and angry world citizens who feel left out. For every disease conquered, a new threat of dangerous toxic waste is uncovered. Technology's faith in progress is haunted by inequities and flaws.

This moral ambivalence was built into the scientific worldview by the idea of a value-free universe, neutral and free of any spiritual force. This view freed humans to exploit and dominate nature at will. It also made the need for human control and dominion more urgent, because lacking a cos-mic, religious meaning that assures a positive destiny, advocates of a strong technological culture need a sense of human command of technology to prevent slipping into destruction, despair, and nihilism. Thus every major technological project carries the burden of providing a hopeful meaning in an empty, alien universe. These implied meanings, whether power, pleasure, adventure, conquest, or cosmic awe, are offered in the unconscious human enchantments of technology.

Seeing the world as value-free is not based on fact but on a metaphysical assumption with some benefits and some dangers, vital to goals of mastery and power over things. The belief in the world as objective and the goal of mastery over nature are essential prior values of the technological worldview. Utopian mythology is an essential part of keeping that vision alive, notably in science fiction that triumphantly portrays the advanced technology of an imagined future, dazzling and/or horrid as it may be. But the ambivalent possibilities of dominion are so uncertain that dystopian myths repeatedly express the fear of loss of control. Natural, technological, and social disasters are disturbing to triumphalism because dominion is not only a value but an article of faith, an enchantment of the dearest kind that must be kept alive for technology-based culture to retain its self-confidence. Technology out of control is frightening. With no faith in cosmic, ontological justice, technological control carries a huge burden of reassurance that its project is workable and right.

Technology and Social Oppression

The increased satisfaction of material needs due to the technological results of dominion has by no means led to the withering away of the will to dominate other people. As William Leiss notes in *The Domination of Nature*, "The domination of man by man has not been rendered superfluous in modern society by the magic of a domination over things" (117). The Iroquois historian John C. Mohawk demonstrates the paradox that, in the pursuit of utopias, people often resort to the most crude behavior: "Utopian ideologies enable plunderers to claim—even to believe—that they are in pursuit of noble goals" (3). Max Horkheimer sees the desire to dominate as the "disease" of reason, growing from its roots: "The disease of reason is that reason was born from man's urge to dominate nature" (177). The technological dream of domination over nature is not socially innocent, for it breeds the enchantment with power that can lead to cruel social oppression.

It is not by chance that triumphal technological enchantment has historically shown great concern with massive power and rational control of that power and one's own behavior but little concern for the souls and ethical cries of oppressed minorities and underdeveloped countries. Technology's will to power tramples too many victims. The general destruction of the environment that technology has promoted, and its limited success in spreading social justice globally, means that domination must shift from a natural to an ethical focus, from machinery to self-restraint. As Leiss writes,

> The idea of mastery of nature must be reinterpreted in such a way that its principal focus is *ethical or moral development* rather than scientific and tech-

nological innovation . . . [for] the gradual self-understanding and self-disciplin-
ing of *human* nature. (*Domination,* 193)

Ethics is not separate from technological innovation, for the will to power is
an ethic. Becoming aware of the enchantments of technology, notably domination, and cultivating the appropriate restraints are stifled obligations. Utopian
triumphalism must beware of the temptation to trample morality by ignoring
its own enchantments.

The New Technological Utopianism

Themes such as equality, communal property, the rise of the middle class, increased cleanliness, health, morality, education, and peace were most common
in utopias until they were overshadowed in the late nineteenth century by the
new genre of science fiction. This new literature championed technological
utopian dreaming but had to add the critical, dark shadow side to the story,
the opposite of utopia's hopes: the dystopian fears of terrible consequences
of technology's powers. Technological utopianism is a major mythology. It is
not simply a separate category of harmless "entertainment," for it expresses
the crucial faiths and fears of the industrial worldview.

Numerous expressions of technological utopianism emerged in the wake of
Bacon's *New Atlantis.* For example, the idea of cultivating new inventions appeared in Tommaso Campanella's *Civitas solis* (The city of the sun, 1602–27).
During the eighteenth century, the motif of inevitable progress creating the
future became a dominant theme and shaped numerous ingenious stories,
beginning in 1770 with Louis Mercier's *An 2440, ou Rêve s'il en fut jamais*
(The year 2440, or a dream if there ever was one; published in an English
translation as *The Year 2500*). The optimistic view that human nature can be
perfected by education was an enchantment promoted by utopians such as
Robert Owen, who wrote in 1813, "'Children are without exception passive
. . . [and] may be ultimately moulded into the very image of rational wishes
and desires'" (qtd. in Emerson 463). The positive impact of new inventions,
inevitable progress, and perfectibility through education have become dogmas
of technological utopia.

By the end of the nineteenth century, during the Gilded Age, industrial
societies gave birth to intensified utopian dreams. They were dazzled by new
technologies: steam engines in trains and boats, telegraphs, machine guns,
dynamite, typewriters, pneumatic tube communications, flying balloons, and
electricity. Some moral improvements were also promising: the official end
of slavery in the United States, the rise of feminism, and socialist ideas of
equality promoted by writers such as Karl Marx and Edward Bellamy. In this

atmosphere, a formulaic utopian journey into the future began with falling asleep or dying and waking up in an astonishing future full of new technologies and an improved society. Three examples, possibly read by the designers of the *Titanic*, illustrate this enchanted genre.

Looking Upward

In 1887 Edward Bellamy published the wildly popular utopia *Looking Backward*, in which a man from 1887 awakes in the year 2000. With no explanation of how human nature has changed or society progressed, the sleeper discovers a polished socialist state, where "the nation guarantees the nurture, education, and the comfortable maintenance of every citizen from the cradle to the grave" (70). Greed and egotism have mysteriously dissolved into the "rationality" of patriotism (207) and willing service to others. The elite classes of the Gilded Age and the masses of filthy poor have been eliminated by the magical suppression of the quest for power and wealth. Bellamy imagines an "illimitable vista of progress" (130), confident in the ability of science to explain former mysteries (140). Jails and lawyers are eliminated, for lying and crime have faded into oblivion with brutishness and wasteful luxury. Rather than money, each citizen is issued a "credit card," or a pasteboard card with a set value that is punched with each purchase. Society is organized as an "industrial army," including women, with a hierarchical order eliminating individualist competition. Bellamy is primarily interested in the economic means of bringing equality and skims over most technological details, but he leaves the reader with a dystopian shiver with his term "industrial army," since, within a generation, the first event of the twentieth century to shatter such utopian enchantments was the horribly destructive industrialized armies of World War I.

But before that debacle, Byron Brooks published *Earth Revisited* in 1893, in which a wealthy Brooklynite dies and awakes in 1993 to a Brooklyn transformed beyond his dreams. Streets clean of smelly horse droppings are filled with electric carriages charged with solar energy, and speedy winged dirigibles ("anemons") are docked on the roofs of houses. Pneumatic tubes whisk meals and clean clothes to homes from community dining halls and laundries. Public schools surpass churches in number, and communication with beings on Mars is common, while property inheritance is abolished to assure democracy. Women are generally considered equal to men, and the "harmonic telegraph" hastens communication between electric typewriters. Reincarnation is a common belief, and religious creeds are replaced by a pragmatic ethic, freed of any hint of sin, that is cheerfully practiced by the populace. Theology is hastily collapsed into science with the saying, "Real Science is true religion" (122).

But these optimistic enchantments were accompanied by darker forebodings. H. G. Wells, along with Jules Verne, invented the genre of science fiction. Wells introduced the enduring fantasy of the Time Machine in his 1895 novel by that name, but he challenged the notion of inevitable progress by showing a declining future. Wells again challenged the optimistic utopian formula with his 1898 novel *War of the Worlds*. Closer to real earthly issues was his 1899 book *When the Sleeper Awakes*. Wells's insomniac Sleeper awakens two hundred years later in a glass case in a new London of thirty-three million residents. While he was asleep, his fortune multiplied incredibly to make him "Master of the World" and thus a threat to the current establishment. Struggles ensue between a ruling council and the working masses, who see the Sleeper as a resurrected savior who has come to liberate them.

Along the way to discovering his exalted identity, he discovers new technologies such as huge buildings, large ceiling ventilator "wind-vanes," cylindrical storage of books and music, the moving picture machine, called a "kinetoscope," which he judges to be a "modern replacement of the camera obscura" (142). "Babble machines" fill public areas, spouting propaganda. Wells has little faith in Bellamy's happy utopian industrial army, for London is filled with an unhappy, uniformed working class, and an army of Africans is brought in airplanes by a treacherous leader to subdue the masses. The Sleeper had hoped that the future would bring a "free and equal manhood," but,

> [a]fter two hundred years, he knew, greater than ever, grown with the city to gigantic proportions, were poverty and helpless labour and all the sorrows of his time. . . . He had heard now of the moral decay that had followed the collapse of supernatural religion in the minds of ignoble man, the decline of public honour, the ascendancy of wealth. For men who had lost their belief in God had still kept their faith in property, and wealth ruled a venial world. (169–70)

The Sleeper, endangered by the web of malicious politics into which he awakes, escapes in an "aëropile."

Wells recycles in this tale the familiar Rip Van Winkle literary device of the sleeper awakening into the future. But he rejects Brooks's and Bellamy's utopian dreams in favor of a more cynical dystopia, much as he had smashed hopes of peaceful relations with mythical Martians in his 1898 *War of the Worlds*. Wells is optimistic about future technological advances but skeptical about improvements in morality, human nature, and economic structures. His 1901 novel *First Men on the Moon* offers an exciting technological goal but also warns against the "hypertrophy of the merely rational and scientific . . . power without compassion and efficiency without purpose" (Kumar 4).

The constant optimism in these futuristic novels ignores the major question: whether human nature will improve as well. These fictional technological utopias advance victoriously, while human nature stumbles along with its ancient weaknesses and hostilities intact. Wells became more optimistic as he aged. In *A Modern Utopia* in 1905 and *Men Like Gods* in 1923, he pictured a socialist world state under the benevolent direction of scientists. His optimism was not shaken by the 1912 sinking of the *Titanic*, but perhaps it should have been. Wells, who lived to see World Wars I and II, held on to his faith in the benevolence of science, the technological elimination of toil and poverty, and the possibilities for socialist developments. He expected that mass education would banish superstition and war but concluded that, tragically, people were enslaved by outdated social institutions.

Technological utopianism, even in the face of the technological disasters of the twentieth century, from world wars to ecological peril, has proven to be a robust enchantment with concealed religious themes. Utopias are usually offered as alternative ideologies giving a blueprint for solutions to current problems. They are rarely escapist fantasies. They often reflect simply a dreamy extension of current technologies, such as the old pneumatic tubes. But their technical accuracy or inaccuracy is not the issue here. The unexplored point is their enchantment, their unconscious visions of irresistible, even magical power. Technological utopias are often seen as a panacea, even a gospel or a prophesy, like an ancient religion. As Howard Segal shows in *Technological Utopianism in American Culture,* techno-utopias are often seen explicitly as heaven. One utopian proclaims, "'Eternity is here. We are living in the midst of it,'" while another declares, "'Heaven will be on earth'" (qtd. in Segal 22–23).

Whether explicitly religious or not, the implicit mythology in the background of techno-utopias is the paradisical Garden of Eden, the eternal heavenly land of immortal bliss. This ancient dream has simply taken on a technological garb, preaching faith in the infinite possibilities of technology. This had the effect of inflating humanity's sense of self-importance. The techno-utopians "saw man becoming God, and the engineer his priest" (Segal 90). Of course, this faith seemed justified in the midst of the exciting technical advances in the cleaning up of horse manure on the streets, feeding and educating children, and speeding up travel. At a practical level, these advances were understandably celebrated.

But some critics nevertheless saw the dark side of the rapid mechanization driving the utopian hopes. If techno-utopia is heaven on earth, could there also be a techno-hell on earth? Yes, some said, and they called it dystopia. For the dystopians, technology became not a solution but a problem. Thomas Carlyle criticized the inward effects of the Mechanical Age—the preoccupation with materialism and the dehumanization of society: "Men are grown mechanical in head and heart, as well as in hand" (63).

Similarly, Henry Adams lamented the West's turn from the Virgin Mary to the Electric Dynamo: "Symbol or energy, the Virgin had acted as the greatest force the Western world ever felt, and had drawn man's activities to herself" (1225). Yet Americans felt no goddess as powerful as the steam engine, he grieved, or the electric generator at the Great Exposition of 1900, where even Adams felt the enchantment of this machine pull at him: "[T]o Adams the dynamo became a symbol of infinity. . . . Before the end, he began to pray to it; inherited instinct taught the natural expression of man before a silent and infinite force" (Adams 1222). Adams greatly feared the shift from holy reverence for the divine to pious enchantment with the machine: "All the steam in the world could not, like the Virgin, build Chartres" (Adams 1225). Technological utopianism triumphantly soars with grandiose dreams but struggles with critical technical and moral failures.

That Sinking Feeling

The sinking of the *Titanic* is a dramatic story of a classic enchanted megamistake. It was the first colossal technological disaster of the twentieth century, a case of technological hubris tragically gone sour. It is constantly retold in books and film as a fascinating counter to naïve triumphalism and an expression of fear of the flaws of technological society. The ship's captain, E. J. Smith, in a utopian daze, said:

> "I cannot imagine any condition which would cause a ship to founder. I cannot conceive of any vital disaster happening to this vessel. Modern shipbuilding has gone beyond that." (qtd. in Spignesi 18)

These are the proud words of the pre–World War I Gilded Age, an age so enchanted by technology's successes that it was confident in "progress" and boastfully triumphalist about its latest achievements. Perhaps it is understandable that the successes of the steam engine, electric lights and motors, wireless communication, photography, repeating weapons, dynamite, the airplane, and the automobile would push many beyond confidence and hubris into a triumphal captivation with their invincibility. But this illusion wrecked common sense and professional judgment and sank the majestic "unsinkable" *Titanic*.

The *Titanic* was the largest and most luxurious liner of its day, the largest moving thing ever made. She was eleven stories high and four city blocks long (Lord, *Remember*, 174). But, unknown to many, this grand monument also represented the consummation of a fifty-year decline in safety standards for ships. In the name of pressures to be economically competitive, large ships like this actually reduced a number of proven safety factors, such as watertight compartments, lifeboats, and double hulls. When it departed from Southampton in April 1912, carrying 1,316 passengers and 891 crew, the *Titanic* was a proud, competitive ship, racing across the Atlantic, about to become an embarrassment to conscientious sailors and the twentieth century's first tragic symbol of the utopian enchantment of technology.

Some earlier ships had better safety measures built in. In 1862, the *Great Eastern* scraped a huge hole—nine feet wide and eighty-three feet long—in its outer hull on an uncharted rock off Long Island, which was far worse than the damage to the *Titanic.* But it had a double hull, so its undamaged inner hull enabled the engine room to remain dry, and it made it to New York Harbor with no loss of life. But by 1912 the *Titanic* had diminished the double hull to only a double bottom and reduced the extent of watertight compartments in the name of space and luxurious items, such as the grand staircase (Lord, *Lives,* 22). Wild optimism and luxuries for wealthy passengers were deadly charms built into the *Titanic.*

Why was this ship sailing so far north in Atlantic waters in April? Six wireless warnings of icebergs were received but ignored by the *Titanic* on the Sunday before it sank. The water temperature had plunged below freezing that night. Why were the lookouts not provided with forward lights and binoculars? Why was the ship speeding through such treacherous waters at 22.5 knots (Lord, *Remember,* 2), while other ships, such as the nearby *Californian,* had stopped in order to avoid hitting ice (but ignored the *Titanic's* emergency flares) (101)? These megamistakes were driven by a triumphalist enchantment with technology. As the proud ship hit ice and sank into the freezing water at 2:20 A.M., the horrible freezing deaths of about fifteen hundred souls (176) was ensured by the cocky collective mystique among some of the finest operators of the latest marine technology. At dawn, rescue ships were astonished to see a solid sheet of ice nearby and huge icebergs as far as the eye could see (147).

Our dreamy optimistic fascination with technology is so ambiguous, it seems, that the story of both the proud achievement and the fearful failure of this ship continues to awaken pride and fear, expressed in numerous books and films. But by no means did that ship's tragic sinking stop our intriguing trance. Its lessons may have saved lives, but the trance is now worse. In 1955 Walter Lord concluded,

> Overriding everything else, the *Titanic* also marked the end of a general feeling of confidence. . . . Scores of ministers preached that the *Titanic* was a

heaven-sent lesson to awaken people from their complacency, to punish them for top-heavy faith in material progress. If it was a lesson, it worked—people have never been sure of anything since. (Lord, *Remember*, 112–13)

But he was wrong. The twentieth century's triumphal faith in material progress blithely continues. The century's disastrous technological wars and wrecks began again within two years of the 1912 sinking. The enchantment of technology even intensified its blind, brute energy. This dream is so essential to motivating industrial society that many cling to it in the face of its disasters.

The ancient Greeks used the word "titanic" to describe the behavior of the gods of the pre-Olympic tribes; the word included hubris to describe not only pride but wanton insolence, insult, violence, and outrage. Titanic triumphalism is the aspect of technological enchantment that cultivates a particular pride and insolence, which leads to blindness and disaster. Pride in the fact that new machines can be built that are bigger and more powerful than ever bloats the collective ego of industrial society. Competition to be the first to achieve such success multiplies the effect. Insolent disregard for the consequences accumulates, and tragic disasters ensue.

In all the excitement of the launching of the great new *Titanic* in Ireland, the editor of the *Irish News and Belfast Morning News* raised the impertinent question of why this new ship was named after a tribe of gods who "'symbolized the vain efforts of mere strength to resist the ordinances of the more "civilized" order established by Zeus, their triumphant enemy.'" He concluded tamely that the owners knew best (qtd. in Lord, *Lives*, 15–16). But on that horrifying night in 1912, it was the ancient Titan Oceanus who won the battle with the proud new industrial enchantment.

What kind of people and culture have we become that we remain so blind to these problems? Are we so inflated by the new accomplishments of our technology that we think we can easily triumph over the larger powers of nature, such as the ocean's treacherous ice and vast depth? One clue is provided by the failed attempt in 1996 to raise a "big piece" of the sunken *Titanic* with balloon floats. These fantasies totally neglected the massive weight of the wreck and the powerful effect of the rough seas (Spignesi 318–21). Here again was the stubborn *Titanic* enchantment, the proud, utopian denial, fantasizing that it could perhaps somehow redeem the blind optimism by resurrecting a rusty remain.

In the 1979 film *SOS Titanic*, a grieving widow, whose husband went down with the *Titanic*, is rescued and safely heading to New York on the *Carpathia*. Another woman attempts to comfort her and others by affirming faith in God. The new widow replies bitterly with the film's final line: "God went down with the *Titanic*." No one can doubt the pain and bitterness of all those survivors, but now we must ask, "What kind of God went down with the *Titanic*?" Was the biblical God outweighed by that tragic pain?

In *The New Polytheism: Rebirth of the Gods and Goddesses*, David L. Miller shows that while western culture is consciously monotheistic in its biblical faith, in actuality, under the surface or even on the surface, we find many other godlike forces, and the very name "Titanic" symbolizes that theme. Greek gods have long fascinated European and American thinkers, but Miller brings them out of the museums into the realm of the contemporary collective unconscious:

> It is not that we worship many Gods and Goddesses (e.g. money, sex, power, and so on); it is rather that the Gods and Goddesses live through our psychic structures. They are given in the fundamental nature of our being, and they manifest themselves always in our behaviors. The Gods grab us, and we play out their stories. (Miller 59)

From this point of view, the *Titanic* tragedy does not mean that the biblical God is dead but that other gods or psychic dynamics were also at work. The titanic powers in human nature still seek grandiose power and pleasure and still strive to master Oceanus with huge machinery. Titanism still offers grandiose illusions of omnipotence.

The large-scale triumphs of industrial technology are worshiped at the altars of Prometheus, who stole fire from the gods, and Hephaestus, the divine blacksmith-technician-engineer. Miller claims that Hercules the hulking hero is at work in the brute force of modern technology and the heroic stances of its builders and operators. Pan, the god of panic and irrational violence, was at work that sinking night. Zeus, the god of kings, leadership, and aristocratic, royal authority, oozes out of the White Star owners' and the captain's imperious neglect of iceberg warnings. Cronos lurks in the conservative traditions of upper-crust European culture that suppressed the lower-class passengers and neglected the importance of the new wireless telegraphy that warned them of icebergs. These archetypal psychic forces are not simply "pagan" texts and statues safely relegated to musty tomes and museums; they are active forces perennially struggling under the dominant monotheism of consciousness.

These lingering old gods are examples of the enchantments remaining in the very cultural world that analytical thinkers had proclaimed disenchanted to free them to build their machines undistracted. Trying to articulate these fascinations that get built into our machines, such as the great *Titanic* turns us to mythic language. Most important, examining the enchantment of technology can free us from the triumphalist illusions that pretend to be in control and deny the clouds of mysterious motivations that actually give reason its vitality.

The subject/object metaphysic of modernism also failed that tragic night. All the calculations that went into building the *Titanic* were driven by triumphalist dreams summarized in the fantasy of unsinkability. "Objective" judgment flew out the door from the day when a committee decided on a small number of lifeboats, and it flew farther as the pride of the crew persisted with the race despite several warnings of dangerous icebergs. It is inadequate to imagine that subjective judgments overcame rational analysis. Rather, the ship itself was an incarnation of the fantasy of titanism, the triumphalist optimism that this unsinkable ship could race across the Atlantic with its severely diminished safety features. The ship itself was not only an object in a mechanical world. It was arrogant triumphalism built in steel, recklessness afloat.

Steered by a proud, aristocratic illusion of superiority, the hands and minds that interacted with the machinery aboard formed a doomed man-machine. These unified wholes come from a deeper place than subjects separated from objects. Ironically, the attitudes guiding the construction of the ship and abetting its collision with an iceberg were colossal mistakes, and sufficient attitude adjustments have yet to be made.

An even more dangerous failure was the 2001 sinking of the twenty-four-thousand-ton, 118–man Russian nuclear submarine *Kursk,* because of the threat of the spread of its nuclear radiation through the world's oceans. Fortunately, the *Kursk* was raised by an international effort before it contaminated the ocean, but at immense expense (Vizard 42–47). Utopian triumphalism is shadowed by dystopia, and naïve optimism coupled with more powerful technology increases the power of danger.

Futurist Predictions

Futurist predictions still fill the pages of popular technomagazines, but they are slowly adding more ecological technologies such as solar and wind electrical generation and hybrid and fuel-cell cars alongside the continuing enchantment with high-speed, gas-guzzling, polluting vehicles. Enchantments with technology continue to promote mechanistic visions at a rapidly edited pace, usually in untempered triumphalist form. Occasionally, predictions are tempered by social-justice, ecological, and feminist agendas, but they still tend to strongly press forward the agenda of industrial society in highly imaginative forms (Kumar 16–17). Certainly societies need visions of hope, but ours are too heavily technological.

Futurist technological predictions, common among utopians, are a particularly slippery undertaking, as the variety of positive and negative outcomes reveals. Only about one in four comes true (Schnarrs 32). Unfortunately, the old adage "If God had wanted us to fly, He would have given us wings" has had the edge for most of history. Mistaken denials, for example, may refuse to imagine technologies that later blossom, such as personal computers, which IBM famously ignored until the Apple Macintosh surprised them and others with its appeal. The popularity of the Internet was also unforeseen. Now utopians must be embarrassed by the crude commercial, sexual, and criminal uses of computers.

But old-fashioned skepticism is losing ground to techno-optimism. Mark Twain was skeptical about the value of telephones, but he was wrong. The human desire to communicate has a huge appetite, as the e-mail and cellphone crazes illustrate. The insatiable urge to communicate is often highly emotional and may have little to do with information. It has a large enchantment component, including the quest for love.

The urge to fly is another insatiable human desire. The speed of travel alone is enchanting, and the wonder of the power of jet airplanes and rockets is a further passion. The moral ego's yearning to touch the ineffable transcendent mysteries of the heavens and to escape the earth's practical concerns is an old utopian dream, symbolized by Icarus, Bellerophon, and other mythical legends. Technology was fated to be put into the service of the yearning to touch the sky.

Other wide-eyed mistaken predictions have not come true yet, such as nuclear home furnaces. These show the inflated dreams of overly optimistic technological enchantments whose practical application did not work. In 1979, the idea that techno-utopianism seemed to be fading was reflected in Sam Love's article, "Whatever Became of the Predicted Effortless World?" He points to a few indicative mistaken predictions: the 1920s and 1930s gave birth to numerous predictions of robot-prepared meals, disposable clothing, personal helicopters, cars commonly driving at a hundred miles per hour, nuclear-powered vehicles, and radar-guided automotive steering. Some of these have taken limited form (with ecological problems), such as disposable diapers and nuclear-powered submarines. Most have not taken practical form.

In 1939, David Sarnoff, president of RCA, naïvely predicted that "'television drama of high caliber and produced by first-rate artists will materially raise the level of dramatic taste of the nation, just as aural broadcasting has raised the level of musical appreciation'" (qtd. in Schnarrs 53). This is another example of how inventors or early promoters of a new technology may have absolutely no concept of its actual use because they are enchanted by either narrow views or utopian visions.

Those involved in post-Hiroshima energy production were thoroughly enchanted with nuclear power, which seemed to be an immense and safe source of energy. R. M. Langer predicted "'a power plant the size of a type-writer,' containing a pound of uranium." In even bolder strokes, he wrote, "'We can look forward to universal comfort, practically free transportation, and unlimited supplies of materials'" (qtd. in Love 88). These triumphalist enchantments reveal the familiar obsession with unlimited power and blind-ness to the side-effects, such as safety hazards and ecological disaster.

Similarly, the bottomless human desire for pleasure lurks behind predic-tions of robot-prepared meals, high-speed cars, personal helicopters, and radar-guided vehicles. High-speed cars have succeeded in fulfilling their en-chantment, despite their widely ignored problems. The radiation dangers of nuclear power have brought strong and controversial responses from those who understand it as an obvious carcinogen, but this has not deterred those enchanted by the dream of unlimited power, who want to build still more nuclear power plants.

Our task is not to debate the reliability of futurism but to explore its influ-ential imaginative enchantments. One of these is the obvious truth that "today's science fiction is often tomorrow's science fact," because we *are* flying in space and we *do* have robots and computers. But this should not lead us to neglect mistaken predictions and their mythic elements. Today's science fiction is often *not* tomorrow's fact but today's utopias. We do not have colonies on the moon or disarmament. Steven Schnarrs outlines salient examples in his book *Megamistakes: Forecasting and the Myth of Rapid Technological Change.*

Of course, history is littered with erroneous predictions, and it is easy to laugh at them afterward with triumphal glee. But prediction itself is an en-chanted expectation. How many men in the United States in 1890 thought that women would be voting in 1920? Who in 1975 would have ever dreamed that by 1995 recycling would be a widespread phenomenon? Will nuclear energy or solar, wind, and water energy production make more gains? Both rational analysis and enchantments are involved in predictions. Taking account of the enchantments is a neglected art.

The most influential enchantment in futurism is optimism: "most tech-nological forecasts are far too optimistic. Optimism results from being enam-ored of technological wonder" (Scharrs 47). One of enchantment's enemies, down-to-earth cost-benefit analysis, shows the price-performance failures of some predictions, such as rear-view TV cameras in cars and human space travel (Schnarrs 85). There is obviously a need for creative vision but not na-ïve utopian dreams of paradise or self-indulgent pleasures for the rich. The solution is to be imaginative and creative but avoid technological wonder. It is dangerous to fall in love with technology, no matter how popular the latest grand utopian visions and the feverish, triumphalist excitement.

While technological utopias play a valid role in inspiring creative solutions to cultural problems, the greatest downside is that focus on technological solutions may overshadow humanistic solutions. The danger of utopian technological dreams, however, is not only that they tend to give permission for dreamers to undertake massively expensive technological projects and dangerous, mistaken undertakings but that quixotic triumphalism too often gives ordinary people permission to ignore ethical restraints and inflict unforeseen damages and injustices that they would not have tolerated without their utopian enchantment.

Too often we enter a triumphalist technological trance of utopian visions that drags good judgment into a fog and invites aggressive, self-righteous behavior in the name of some dreamy ideology, such as progress or the conquest of space. Destructive technology is often too hastily organized to serve the defense of investments, deep resentments, injured pride, or despair that trigger a utopian fantasy justifying horrendous acts, such as the militarization of space. No technology involved in such programs can be considered a neutral tool. Technology is built and used for highly political purposes. Technological triumphalism is a titanic surge of enchanted modern consciousness unabashedly engaging in dreamy or dangerous fantasies and magical transcendental visions.

6 The Space Cowboy

From the beginning, the space race was less about science than about technological display.

—William Stahl, *God and the Chip: Religion and the Culture of Technology*

The Six-Shootin' Cowboy

In Stanley Kubrick's classic 1964 film *Dr. Strangelove,* the pilot of the ill-fated B-52 that cannot be stopped from dropping its nuclear bomb is named "Tex." He connects the enduring mythic cowboy of American legend with the high-tech world of military bombers and their apocalyptic missions. This darkly comic warrior fulfills his fate by suicidally riding a nuclear bomb down onto a Soviet base waving his cowboy hat, yelling "yahoo!" This tragic image satirizes the enchantment of the cowboy spirit in the history of North American aviation and space technology, the space cowboy who braves new frontiers with the "right stuff," driving the latest fast, hot, and deadly technology to the edge of frontier-space and the edge of catastrophe.

The western cowboy is a mythical creature, made not by cows but by technology. He means far more for North American culture than a realistic western plains rancher herding cattle. For the most part, the cowboy is a fantastic, larger-than-life archetypal hero. Part Jesse James, part John Wayne, and part Matt Dillon, this legendary character, originating in industrial society's nineteenth-century mythology, was made mythic because of the gun, a newly mass-produced repeating pistol. What would a mythic cowboy be without his six-shootin' iron strapped to his side? Nothing but a humble cow herder.

The cowboy was a key starting point for the enchantment of technology in North America because of gunpowder and the power it symbolizes: conquest of natives and acquisition of a vast land. Of course, the gun also symbolizes crude phallic power over the more subtle but beautiful forces of the feminine and native Mother Earth. The gun amplified the power of hunters whose lives depended on a little meat as well as the children 150 years later shooting other children in schools. The "cow-boy" plays a leading role in the historic/symbolic drama of the archetypal Wild West pioneer by pushing forward the European immigrants' "frontier."

Of course, contemporary U.S. patriotism requires that we remember only the finer side of the cowboy's adventures and elevate them to heroic status. After all, manhood in the United States is heavily invested in the image of the rugged, stone-faced individual, supplemented by strong horsepower and wielding guns, ready for a fight at any time and playing his part in taking over native land so it can be exploited ("developed") by industrial society's ravenous desires for meat and gold. This enchantment lingers in the widespread all-American denim jeans, which were created out of rugged tent canvas for the gold miners. Wearing those jeans, we imaginatively step into cowboy mythology on the frontier, not only displaying a youthful bit of bohemian style but riding in the costume of that horsepower, shooting that gun, rushing for that gold. Those pants are not simply objects in a neutral space but strong images of the western past that still offer identity dreams, a costume for a role, like cowboy boots and hats worn by truck drivers.

The cowboy myth has faded with the old John Wayne movies, but the heroic impulse and its North American technological version has simply taken on new forms with each new technology. The cowboy enchantment survives in the mythic airplane pilot and the astronauts with the "right stuff," each pressing the frontier-space of technology outward. The mythic cowboy is not about horses and leather but machines, weapons, and the expansion of industrial society, where even ranching feeds the meat "industry."

Leaping toward the Infinite

The airplane's leap into the sky wildly expanded the world's imagination of technological possibilities. It was the long-awaited fulfillment of the age-old dream of flying among the clouds and the stars. The remarkable achievements of the Wright brothers, Charles Lindbergh, and Amelia Earhart echoed the heroic enchantment of the cowboy pushing the frontier with a brave new machine. Successful aeronautical engineering has been triumphant in powerful ways. But more subtle psychological and spiritual forces drift through the air and the soul in the enchantments of flight.

Ascension can be an erotic, seductive elevation, a voluptuous refinement of desire. Rising above the earth's heaviness is an ecstatic jump to new heights, a wandering in a vast openness, an expansion of horizons and consciousness. Overcoming our gravitational mass is hopeful, promising a new future, a deliverance to a new world and growth. Flying is lightly mounting on angel's wings, soaring toward the infinite, yearning for ontological visions of heavenly expansiveness into the beyond. Taking off brings out fearless courage to leap past the ordinary into the transcendent endlessness of space. It offers vastly expanded spectacles of reality—seemingly ultimate reality—in outer space.

Without such boldness, of course, flight can be fearful, as gravity and stability exert their downward pull. Falling is the opposite of flight, and as Icarus and other images tell us, preparation and caution are required to avoid the deadly excesses of recklessness that flight offers. Clouds suggest this ambivalence, for in their light, sunny form they are marvelous mountains of irresponsible drifting, shape-shifting transformation. But in darkness blowing fierce winds, they can obscure and rumble, pour moisture, and bring terrifying and destructive thunder and lightning. As Gaston Bachelard says in *L'Air et les songes* (Air and dreams), wind also has a swift multivalence, as it flows across the globe gently or furiously: "All the phases of wind have their psychology. Wind can be exciting or discouraging. It roars and it complains. It passes from violence to grief" (262).

The transcendent presence of flight is evoked by its wonder, its majestic views, and its light. The diurnal rotation of the earth brings the sunrise and sunset that evoke a sense of the Eternal Return, the cosmic cycles of existence that frame our being. Bachelard sees the night's constellations as a screen for unconscious projection of collective dramas: "The Zodiac is the Rorschach test of the infant humanity" (*L'Air* 202). Telescopes can reach out to the vast, sparkling nebulae's "spasmodic chaos of incandescence. . . . In the nebulae of creation, night meditates silently, the primordial swarms slowly assembling themselves" (229). I too am enchanted when the stars transform my gaze from the flat impression of the sky into the depths of space. I try to imagine the vast distance of the stars, to caress the spherical moon with my eyes, feel its roundness, and wonder at its suspension in the void by earth's gravity.

The very word "spiritual," like the Greek *pneuma* and Latin *spiritus*, refers to the invisible breath of life and its transcendent source above, as distinct from embodied soul and earthly reality. Flight offers a spectacular, ecstatic, dreamy, and dangerous evasion of down-to-earth, practical concerns, breeding a utopian quest for a return to paradise. The sensational machines of flight are, as Jean Brun says,

a daughter of dreams and a child of the impossible. From the wings of Icarus to those of the airplane, in the fullness of time is unfolded the harmonic of the sigh of Goethe: "Ah! How I wish to throw myself into endless space, to float above the enormous abysses." (*Masques* 51)

The will to fly also elevates the will to dominate:

> From the wing of the bird to that of the airplane, man's desire to pull up
> anchors from earth becomes concrete and distorted in order to achieve the
> ecstasy that seems promised to him from those heights from which one domi-
> nates the world. (53)

Evasion and domination are moral shadows of flight, a danger of "flightiness"
that can lead to disaster on earth as well as in the air. "Top gun" fantasies of
military pilots as domineering warriors cultivate not only aggressiveness but
recklessness.

Sky mythology also opens to visions of divinity. In his classic *Patterns in
Comparative Religion,* Mircea Eliade says,

> There is an almost universal belief in a celestial divine being who created the
> universe and guarantees the fecundity of the earth . . . the sky itself directly
> reveals a transcendence, a power and a holiness. . . . The sky shows itself as it
> really is: infinite, transcendent . . . 'quite apart' from the tiny thing that is man
> and his span of life. (38–39)

The infinite heights beyond human reach express the cosmic majesty of absolute reality, everlastingness. The stars, for some Australian Aborigines, are
the campfires of the ancestor spirits. Native American words for "sacred,"
such as the Iroquois *oki* and Sioux *wakan,* are linguistically very close to their
words for "on high." Of course, the biblical God is also traditionally pictured
as a sky deity inhabiting a remote, cloudy heaven.

Ancient accounts of flight combine possible technical accomplishments
with mythology. Kites were known in Asia long ago, sometimes forty feet or
more in size—large enough, it was said, to lift a person up. Ancient legends
and art repeatedly tell of the worldwide yearning to fly, whether lifted on a
Greek chariot, an Italian balloon, a Babylonian eagle, a Chinese dragon, an
Arabic flying carpet, or Taoist magical shoes. By the nineteenth century, kite
makers in Europe succeeded in lifting people over a hundred feet up with
several connected kites (Laufer 31–43).

Modern atheists who put their faith in technological progress have not
really abandoned sky mythology. When balloons and airplanes began to allow humans to edge upward into the heretofore inaccessible blue heights,
sky myths continued to mingle with technological development in significant
ways. Sky mythology is not simply imaginative icing on the practical realities
of technological history; its images and fantasies motivate the development
of technological systems, and we find them behind airplanes and rockets. Far
from rejecting myth, sky technologies embrace its dreams and fly machines in

search of those dreams. But these fascinations are highly ambiguous. A Greek myth long ago told of the dangers of the sky enchantment.

The Icarus Enchantment

In ancient Greek myth, Icarus was the son of Daedalus, the clever inventor. Daedalus built the labyrinth for King Minos in Crete, where the legendary Minotaur is imprisoned. Later, however, the angry Minos imprisons Daedalus and Icarus in their own labyrinth. Freed by Minos's daughter Ariadne, Daedalus makes for himself and his son wings of feathers and wax so they can escape the island before Minos catches them. As they prepare to fly, the father warns his son not to fly too low, lest the sea splash his wings and weigh him down. Nor should he fly too high, for the sun will scorch them. Above all, Daedalus warns his beloved son to follow him closely and not be diverted by the celestial creatures above. Taking off, they pass the islands of Delos, Paros, and Samos safely. But the daring Icarus becomes overconfident as he soars through the air, neglecting his father's admonitions. Thrilled with his newfound power to soar high above the Mediterranean, he flies too high, so near the sun that he scorches his feathery wings, and the wax melts. Terrified, he plummets into the sea (Ovid 187–90).

This ancient myth portrays an ageless psychological pattern: the archetypal constellation of flying and falling. Flying can be symbolic of romantic, utopian visions of ambitious achievement, and this was so even when real flying was impossible. Flying evokes the tendency to strive for the highest goals imaginable, to "reach for the stars" or stretch one's limits to the outermost. Flight for Icarus and Daedalus is an escape from imprisonment. The enchantment of escape from problems is characteristic of the adventurous wanderer, like the adolescent cowboy or pilot, heading off into the sunset, leaving behind the responsibilities of a family. Flying also implies the mental capacity to imagine richly and think philosophically, envisioning the "big picture" as from above, seeing the farthest horizons and stretching the mind to reach beyond. Icarus symbolizes the age-old yearning for *apotheosis*—becoming godlike, omnipotent, omniscient, surveying the world from above, the ultimate creator and savior. Flying is approaching the infinite. The glory of technological flight easily embraces the archetypal Icarus enchantment.

As every real pilot now knows, however, the Icarian excitement must be tempered by caution and serious safety measures, because the shadow side of the glory of flight is the tragic pain of falling and crashing. This is the Icarus enchantment. If flight approaches the enchantment of Paradise, the fall echoes the primordial tragic flaw, that biblical Fall and expulsion from Eden. This shadowy threat lurks throughout the history of aviation and space flight. The enchantment of flight is always torn between the romantic, utopian dream of "getting high" and the terror of the fall.

Early dreams and experiments with flying machines long ago enacted the Icarus enchantment. Leonardo sketched some impossibly heavy and small flying machines. The nineteenth century was filled with determined experimenters crashing overweight wings and balloons. Though balloon technique succeeded first, the dream of widespread use of dirigibles went down in flames in 1937, when the German zeppelin *Hindenberg* exploded.

The dream of human flight soared in less than a century from the Wright brothers' 1903 flight, through Lindberg's 1927 trans-Atlantic flight and Amelia Earhart's bold 1937 effort to fly around the world, to jumbo jets and space exploration. The magic of flight took the collective imagination up into many flighty dreams. As a boy in the 1950s, I remember the utopian enchantment that soon everyone would have their own personal helicopter. I don't know about you, but I did not get mine. Of course, I could now buy a helicopter for a small fortune, earn a helicopter pilot's license for another large bill, and pay exorbitant insurance fees. Should we wonder why we don't see so many in the skies, which would be very crowded and dangerous anyway? Aviation's history is filled with the ambiguous Icarus enchantment—its expensive, evasive dreams, its remarkable successes, and its terrible disasters.

The Wright Wizardry

Orville and Wilbur Wright finally fulfilled the dreams of centuries of tinkerers striving to invent the machine that would free humanity from gravity and let us soar like the birds. They figured out the basic techniques of controlled, powered flight. Raised in Dayton, Ohio, the two sons of a church bishop were mostly self-taught. But with their outstanding mechanical talents, they arrived on the stage of history at the right moment, because aerodynamic theory, structural engineering, engine design, and fuel technology were all emerging. All they had to do was put the pieces together. And they brought the right temperament to the task: they were obsessed with the dream of flight but also sensible enough to be cautious and practical, to use mathematical analysis and avoid killing themselves. Some of their predecessors and competitors offered interesting contrasts. In Germany, Otto Lilienthal had successfully flown what we would call a hang glider, but he died in a crash in 1896. That same year in Virginia, Samuel Langley had flown an unmanned steam-powered glider. The race to discover the secrets of flight captivated many.

Wilbur and Orville Wright analyzed the key elements of flight control into the three axes of motion they saw in birds: pitch (climbing or descending), achieved with a forward elevator; yaw (turning right or left), achieved with a rear rudder; and roll (rotating right or left), achieved with "wing warping." They were the first to master control of roll. Their new movable pilot controls, with flexible flaps, were a major contribution to aerodynamics. They also carefully improved the wing camber. They built three gliders to improve these techniques in 1900, 1901, and 1902, at the windy beaches of Kitty Hawk and Kill Devil Hills, North Carolina, where a replica of their successful biplane is displayed today.

But their success quickly attracted other enchanted competitors and their egos. One was Octave Chanute, who often helped them. He believed in a team approach and so invited himself and others to observe the Wrights at Kitty Hawk. However, Chanute eventually turned on them and tried to steal their glory by calling them his "'devoted collaborators'" who were carrying his work forward to completion (qtd. in Crouch 251). The airplane was not simply an experimental machine, a neutral, utilitarian object in space. It attracted pride and competition for the glory of conquering the skies.

Augustus Herring was another competitor, a proud, self-confident man who thought of himself as the leading American aviation experimenter. After years of struggling, he visited Kitty Hawk and was humiliated by what he saw the Wrights doing. So he tried to get in on their success by offering a three-way partnership, but they refused (Crouch 75, 239). After that, when Samuel Langley asked to visit Kitty Hawk, they also turned him down (Crouch 240). The passionate obsession of inventors to be the first to fly was becoming intense.

These early planes show how machines reflect personalities. For example, some early airplane experiments were elaborate, dreamy, awkward flops with flapping birdlike wings or whirling discs. Chanute was a civil engineer, and his hang gliders expressed the rigidity of a bridge, but the Wrights' design was simple and mathematically planned. They were bicycle builders, and their planes expressed the light weight and control of a bike. Having solved the basic control problems, they designed and made their own engine and propeller. The jaded Patent Office rejected their first two applications for a powered, controllable airplane patent in 1903, seeing them as the latest in a long line of crank tinkerers (Crouch 246).

On December 8, 1903, Samuel Langley's final effort at flying failed as his poorly designed machine flopped into the Potomac, and his wet, cold pilot had to be rescued, cursing angrily. Just nine days later, at Kitty Hawk on December 17, Orville Wright flew the first successful controlled, powered flight, even though it lasted only twelve seconds and went a mere 120 feet. Later that day Wilbur flew the best record yet: 852 feet in fifty-nine seconds.

The correctness of their design was demonstrated beyond a doubt. One of the excited observing locals ran back toward Kitty Hawk shouting, "They done it! They done it! Dam'd if they ain't flew!" Some newspapers ignored the event, while others garbled and elaborated on it shamelessly, reporting that they flew three miles, sixty feet up (Crouch 262–71).

Early applications of flight were military, showing that the enchantment of flying had an early will to power and warrior spirit. But, stung by the failure of Langley's attempts, which cost the U.S. War Department fifty thousand dollars, the military refused to acknowledge that the Wrights had successfully mastered flight. Suspicious of copycat rivals, the Wrights declined to demonstrate publically without a major contract to buy their invention. This delayed acceptance until their American and European patents were finally approved between 1904 and 1906, while their rivals in France—Voisin, Blériot, and Santos—came close to duplicating their achievement. These bold inventors were informed by photos of earlier planes but did not understand the control and power principles that the Wrights had grasped. The race aspect of enchantment was heating up. Who would be the first to get undeniable credit for mastering flight? (Crouch 316–26). The new character of the technological inventor had a lead role on the stage of history by now, and pride of place in a race to be first was a strong, dramatic theme.

Wilbur Wright finally went to France in 1908, where he made the Wrights' first public demonstrations, flying over an hour, breaking all time, distance, and altitude records, and taking up passengers. The excited French, who had thought of the Wrights as "bluffers," realized that they could control maneuvers such as turning far better than their French successors, who had not mastered the roll. The huge crowds at Le Mans and the newspapers were astonished. Wilbur flew for over two hours and won a French prize. "The psychological impact was stunning," comments Tom Crouch: "If man could fly, was any goal beyond his reach?" (369). The enchantment of flight increased in excitement, opening seemingly impossible horizons. Their accomplishment fueled the growing faith in progress. Earthbound for millenia, humans were finally leaping toward infinity!

The U.S. army finally paid attention. Orville Wright made spectacular flights near Washington, D.C., breaking more records. U.S. warplanes went into production by 1909. Enormous glory was poured on the Wright brothers. Banquets, medals, honors, fireworks, and massive celebrations were given to the modest brothers and their gracious sister Katharine. In Washington, President Taft assured them of the nation's pride. In public, the enchantment with flight was bursting with pride and dreams of power and glory. In private, however, a series of painful patent-infringement court battles drained the

Wrights' energy. Although they usually won these court cases against those who sought to steal their achievements, envy over being the first to master flight proved to be one of the most annoying and petty enchantments of the era. Obviously the airplane was not only a useful tool. It was the goal of the race fantasy, growing on large doses of ambition, pride, and profiteering and provoking fierce competition and legal wrangling over patents.

Astonished crowds flocked to spectator-oriented shows where flyers sought to break records once again. At first the Wrights avoided such showmanship, but they were forced into it financially, so they trained a group of young pilots who reached for higher speeds and altitudes. The Dive of Death, a nose-down plunge from a thousand feet, pulling up at the last second, drew mobs of thrill seekers. But six of the nine Wright team pilots ended such stunts with a funeral, so the Wrights' show business ended in 1911. The dark side of the Icarus enchantment was never far away. Yet plenty of spectators who had never seen a plane fly before walked away muttering, "My God, my God!" (Crouch 429–31). The new enchantment with speed and height was a thrill that stretched the collective imagination of what was possible in the new century. In the lap of the enchantment of technology, amazement would soon develop into expectations taken for granted.

The obsession with flight finally achieved the dream of centuries. One of the most exciting technologies ever invented leaped toward infinity. It was on its way to incredible accomplishments in the twentieth century, as well as unimagined destruction. The dream of airpower led to great leaps in the excitement of speedy personal freedom, essential to the American Dream, and also to the horrors of military bombing. In the dark shadow of nationalism's will to power, the double-edged sword of flight began its odyssey. As Laurence Goldstein says,

> The airplane was perceived immediately as a material counterpart of the aspiring spirit, with the enviable ability to fly clear of a creaturely realm characterized by confinement and decay. . . . It provided a mobile and beatific vision of a new heaven and a new earth. (6)

But the shadow of flight would not disappear with that beatific vision, for the lust for ultimate power haunts the power of flight.

The first airplane was borne aloft not only by careful calculation and experiment but also on gusts of freedom, imagination, glory, wealth, pride, competitive racing to be first, ambition, death-defying bravado, a reach for the heavens, escaping earthly limits, and a yearning for ultimate power. The numerous heroic feats of that spellbound journey developed into the aerial dogfights of the First World War and the astonishing first Atlantic crossing by Charles Lindbergh.

The Spirits of Charles Lindbergh

Charles Lindbergh tells of a day of "barnstorming," when he and H. J. Lynch sat in a field in Montana, waiting for a passenger. A large new car drove up, and a wealthy rancher asked for a ride. So Lynch flew him over the nearby town, where he wanted to buzz the main street. Before the pilot could stop him, the cowboy pulled out two pistols and started shooting and laughing, yelling, "I SHOT THIS TOWN UP A'FOOT, AN' I SHOT THIS TOWN UP A'HOSSBACK, AND NOW I SHOT THIS TOWN UP FROM A AIRPLANE' (Lindbergh 431). The American cowboy now had a new machine with which to shoot, to show off, to bully, to recklessly throw around his weight, and to get high from newfound power.

Early aviators delighted in risky stunts that thrilled crowds and fed faith in progress, which would indeed lead to one after another stunning aviation success. Lindbergh was one of the most successful pilots before World War II. He flew air mail between St. Louis and Chicago in 1926, but he also had the daring vision and technical skill to make the first trans-Atlantic flight in 1927. "New York to Paris nonstop!" he dreamed, "If airplanes can do that, there's no limit to aviation's future" (Lindbergh 16). That limitless, reach-for-the-sky dream is part of the enchantment of flight.

But many remained skeptical of the possibilities of aviation. For one thing, it was dangerous, and many pilots did die in crashes. Was flight only another mad dream of H. G. Wells and Jules Verne, a curious Wright brothers' toy, a modest addition to wartime weapons? Or could it really develop into a serious mode of transportation and power? This was a new frontier with many unknowns, but North Americans thrive on the myth of the new frontier, and technological frontiers were increasingly being conquered. So maybe the dreamers were right, and the mysteries of flying could be further uncovered. Lindberg was one of those dreamers. He was determined to expand the potential of aviation. He succeeded, and for his accomplishment he was idolized and tortured.

As he recounts in his autobiography, *The Spirit of St. Louis*, this lanky youth, born in 1902, quit the University of Wisconsin after two years and learned to fly in the army. When he first watched a parachute jump, Lindbergh saw the brave jumper as "a man who by stitches, cloth, and cord, had made himself a god of the sky for those immortal moments" (255). This reach for immortality helped make Lindy such a daredevil. Lindbergh loved adventure and immortality touched with danger (255). He was an excellent pilot, but during the early days of flimsy, underpowered planes and poorly marked night navigation, he had to survive a few bailouts and crashes. When the twenty-five-thousand-dollar Orteig Prize was offered to the first pilot to complete a trans-Atlantic crossing, several prepared for the competition, and Lindbergh's success was due not only to his character but to the tragic and foolish failures of other pilots. His competitors included the well-known Admiral Byrd, who chose to use a three-engine Fokker monoplane.

By contrast, Lindbergh preferred a small and light single-engine plane with himself as the only pilot. Lindbergh was a lonely wanderer from childhood, so he was prepared to fly it solo (Berg 33, 98). As the spring of 1927 approached, several pilots were preparing to make the leap across the unknown Atlantic airways ahead of Lindbergh, but they were delayed by accidents. Byrd's Fokker crashed in New Jersey on April 16, Chamberlain's Bellanca plane crashed on April 24, Davis was killed in a crash on April 26, and Nungesser left Paris on May 8 but crashed and was lost.

Meanwhile, the twenty-five-year-old Lindbergh's plane was hastily built to his specifications in two months. That machine had Charles Lindbergh built into it: no heavy radio, no parachute, a forward fuselage fuel tank, and a light frame with a strong engine. He flew the new *Spirit of St. Louis* east to Roosevelt field, Long Island, where the newspapers were in a frenzy, the airfield was crowded with thousands of onlookers, and the weather was doubtful (Lindbergh 156). Lindbergh, who had been ignored by the press before, was now seen as a contender and dubbed "The Flying Kid" (161). Reporters did not hesitate to make up stories about him, and he complained: "Those fellows must think I'm a cowpuncher just transferred to aviation" (166). The cowboy enchantment was again being transferred to wings. But Charles was not just a kid dreamer; he was also concerned with accuracy: "If pilots are inaccurate, they get lost—sometimes killed. In my profession life itself depends on accuracy" (167). Even while living on the edge, Charles knew how to avoid the deadly Icarus enchantment.

On Friday morning, May 20, after a worry-filled struggle to grab a couple of hours of sleep, Lindbergh headed for his plane, arriving at 3 A.M. While Byrd and Chamberlain were fretting about wet weather, plane tests, and crews, the lanky midwesterner simply topped off his fuel tanks and, weighing two and a half tons, took off from the muddy field at dawn. "No plane ever took off so heavily loaded," he worried. But his well-designed plane hopped along, slowly pulled its weight upward, and cleared some telephone wires by a mere twenty feet. Lindbergh was on his way into the dawn of trans-Atlantic flight—now so common, then so astonishing and dangerous (Lindbergh 183–87). There were numerous unknowns, especially with his dead-reckoning navigation, but, in his wicker seat in that little cockpit with open side windows, he was confident.

Between navigation, avoiding storms, and struggling to stay awake, Charles had many of his 33.5 hours aloft to think. He reminisced about earlier adventures and the threat of death buried in the thrill of flight. Gazing at the stars, he was drawn to the night's immense vista and the transcendent spirit that his journey evoked. Edging toward the stars, he wondered about the meaning of escaping earthly bonds and feeling godlike among the clouds (Lindbergh 301–2). He often mused on the divine mysteries in the depths of the universe. He exclaimed: "What freedom lies in flying! What godlike power it gives to man!" (94).

This sense of godlike power is a strong part of the enchantment of technology, for flying is a huge leap beyond practicality to the edge of the infinite: "It's hard to be an agnostic up here," Lindbergh thought (321). Human technological devices are so frail, and the universe is so vast and infinite. If he were to crash and die, the world would continue on, in an incomprehensible complexity, he mused. We humans are gifted to be conscious of so much, audience of wondrous atoms and stars. What else could it all come from but God (322)?

Visions in the Moonlit Clouds

Lindbergh had stumbled unawares into a totally surprising phenomenon. His daring flight became a technological vision quest, a journey into unknown spiritual horizons. The technological feat of aviation history turned into an archaic visionary enchantment, like a soul quest for initiation into a higher level of consciousness. His sleeplessness, isolation, fasting, and heavenly view transported him much farther than he expected.

Certain Native American tribes send their youth on solitary fasts in isolated locations, often on mountaintops, to meditate and listen for messages from the surrounding spirit world. After a sweat lodge, a few days of isolation, fasting, and a spectacular natural view, the mind is likely to experience a spiritual awakening. Lindbergh had no such intent, but he bumped into this visionary territory during his flight. Unintentionally fasting, he ate nothing until the end and drank little water while alone facing mountainous clouds and the ocean's vastness. Although he was totally unprepared for this vision-inducing environment, he heard from the spirit world and was awakened to a mystical awareness. He repeatedly fell into a sleepy trance on the edge of dreaming and saw secrets beyond ordinary consciousness (375).

He said he saw unearthly, ghostly presences—transparent, vague forms—passing in and out of the plane, speaking with familiar human voices. These mystical spirit-forms advised him on flight details, reassured him, and conveyed extraordinary messages. He lost his sense of flesh, food, shelter, and life. Free from nature's laws, he almost merged with these vaporous, ethereal beings, maintaining his connection to life by only a thin line that could be crossed at any time. Hovering between earth and these universal emanations of age-old, powerful forces, he felt the spirits to be like old family and friends from past incarnations. Their wondrous presence banished his natural fear of death, transforming it from a fearful end into a wondrous entrance into infinity. Far beyond the ordinary world, "they belong to with the towering thunderheads and the moonlit corridors of sky" (389–90).

Hardly the talk of a hard-nosed engineering-oriented pilot, he sounds more like a primordial shaman, a tribal medicine man who has meditated long and hard in sweat lodges, on vision quests, and has flown on eagles' wings, or a chanting mystic in monastic robes. In the midst of a technological world, Lindbergh's mind opened into an ancient, soaring vista.

Technically inclined thinkers may call these visionary insights "hallucinations," but Lindberg did not lose his ego functions, go mad, and crash. For all his technological orientation, he treated these visions and voices as spiritual guides, as ancient, eternal friends. Hidden in the technical lifeworld, such visionary voices are more than the technical philosophy imagines or wants to permit. Right in the midst of one of the great moments of technological history, out pops an ancient spiritual phenomenon, mysterious visions of spirits that take the pilot to the border of life, to a realm beyond industrial consciousness. The heroic technician Lindbergh even speaks like a Zen practitioner:

> Then what am I—the body substance which I can see with my eyes and feel with my hands? Or am I this realization, this greater understanding which dwells within it, yet expands through the universe outside; a part of all existence, powerless but without need for power; immersed in solitude, yet in contact with all creation? (Lindbergh 353)

Fighting immense fatigue, struggling to stay awake and to maintain consciousness and control of his rather unstable plane, this noble hero of technology journeyed into unexpected space, the soul's space between consciousness and dreamy sleep, through the rational ego of industrial culture into the primordial soul of an ancient paradox of spiritual practices: "My eyes . . . became a part of this third element, this separate mind which is mine and yet is not, this mind both far away in eternity and within the confines of my skull" (362). Like the classic hero returning from a mystical journey, Lindbergh returned unable to articulate the mystical voices in the clouds. The utilitarian land to which he returned had no place for such experiences, and he could not remember what they said (467). This suggests the need for a home for such spiritual powers, a home in harmony with technology. Lindbergh's surprising leap toward eternity took him closer to that home than expected. The enchantment is already in technology's home, even if unnoticed or forgotten.

When he landed at Le Bourget Field in France, the tumultuous reception showered upon the modest Charles Lindbergh by a mob of 150,000 was a riotous outpouring. He was swept up by a human tidal wave madly rushing to lift him up and worship him (Berg 129). Lindbergh had now proven to skeptics that airplanes had a bright future. President Coolidge sent a navy ship to Europe to retrieve him. In Washington, an unprecedented crowd of thousands wildly cheered his modest, brief speeches. New York lavished four days of parades, banquets, and receptions on him, and the *New York Times* devoted its first sixteen pages to his story. Lindbergh struggled thereafter to retain his integrity and guard his privacy, turning down a Hearst movie offer of half a million dollars.

In 1929 Lindbergh married Anne Morrow, who was as shy as he was. They had six children and flew extensively together. Anne became his navigator, radio operator, co-pilot, and a popular writer (Eakin 14). The worst horror of their lives was the outrageous kidnapping and murder of their first child in 1932, a monstrous price to pay for fame. But he loved writing and took fourteen years to carefully compose his Pulitzer Prize–winning *Spirit of St. Louis*, which surprised his audience by describing his visionary encounters. He died in 1974 (Gill 139–40, 197).

Why did Lindbergh's flight ignite such an explosion of excitement, publicity, and tragedy? He certainly did not seek this, for he was committed to developing the potential of aviation and was not interested in fame or fortune. The eruption of this public Lindbergh passion came when he showed to skeptics that aviation did have a remarkable future. The fervor, however, was a collective enchantment with technology, a higher level of captivation with the new power that immensely expanded human horizons and elevated technoculture to new mythic heights. As Laurence Goldstein says:

> The first interpreters of Lindbergh's flight associated his solo exploit with almost every myth in Western culture. They glorified him as a modern hero and his flight as a synthesis of older religions of heaven and modernist faith in the new order of machinery. (9)

This empowerment lit the collective imagination like a wildfire, spreading intense joy to those who could imagine partaking in its bright future and dark threats in the minds of others, which were soon fulfilled in World War II.

The collective trance of flight lit up the landscape of industrial society with new potentials, new horizons, new goals, a new sense of speed, and a wildly enlarged but morally ambiguous sense of power. Lindbergh's "moonlit corridors of the sky" were now opened to major aviation initiatives, offering to earthbound humans the glory of flight once reserved for gods. Aviation's leap toward infinity enlarged the power of the collective ego of industrial society; people gladly grasped the new power. The enchantment of aviation spread this sense of godlike power, as passengers lost their sense of the wonder of flight and began to take it for granted. Numerous aeronautical engineers took on the challenge of building better, faster aircraft. The first jet was flown in 1942. The first successful jetliner was the Boeing 707 in 1954 (Yenne 79). But the history of aviation is not as triumphalist as accounts of successes suggest; it is also full of failed designs, built of misguided enchantments.

The World's Worst Aircraft

Fascination with flight's potential built some of the world's worst aircraft. The obsession with flight leaped beyond reason and built some unbelievable airplanes. Some badly designed airplanes were simply underpowered (Yenne 9). Others were ill-conceived efforts at huge size, such as the 1919 British triple-winged *Tarrant Tabor,* whose top-heavy engines on takeoff tipped it into a hundred-mile-per-hour somersault. Fascination with flying boats reached its absurdity with the Italian *Capronisimo,* a fifty-five-thousand-pound boat with six wings and eight engines, meant to carry a hundred passengers in luxury. On its initial takeoff in 1921, it made it up to sixty feet above the water and disintegrated.

One of the most laughable concoctions was the flying car. As early as 1917 this notion was explored, and some flew, but the hybrid never made a good airplane or car. How would the car carry around the wings and propeller engine? In a trailer. Would the trailer fly along too? No. Though a few prototypes were made and flown, the fantasy of a sky full of these convenient little consumer illusions never took off.

The Northrup B-2 Stealth bomber is the U.S. tailless plane that is made to deflect radar and seem invisible. It looks like Northrup's futuristic flying wing of 1949. Its greatest problem is its cost—more than its weight in gold. In 1988 it passed its flying tests well, but its annual *$8 billion* budget cost more than the entire annual defense budget of most countries in the world. A newer

F-117 Stealth bomber was shot down by Serbian anti-aircraft in 1999, when they saw that the plane's open bomb doors "look like a barn on radar" (Vistica 30). It was an expensive disappointment.

These aircraft were not merely products of bad engineering or financial decisions. They were the embodiments of airy heroic wishes. Well-designed and proven airplanes are the products of good aeronautical engineering, not only enchanted fantasies. The worst craft were not just awkward objects made by ill-advised subjects; they were dreams in wood and steel, fantasies striving to fly.

A Brief Supersonic Affair

Within sixty-six years of Lindbergh's flight, an incredible leap in airspeed made supersonic jets commonplace. Expectations for speed were far higher than ever before, and the thrill of flying captivated the world's imagination even more. But economic factors are considerable, for flying weapons are immensely expensive. After the end of the Cold War in 1989, the Russian air force was so short on funding that they offered thrill rides in MIG fighter jets to civilians. Peter Passell, a writer for the New York Times, paid the Russian air force for the privilege of flying for a few minutes in a supersonic military jet in 1993. After taking off at a forty-five-degree angle, the MIG-29 leveled off at sixty-thousand feet, at the edge of space, hurtling so fast that Passell could not touch the glass canopy because the speed heated it up to two hundred degrees. With two turbojets capable of forty-eight thousand pounds of thrust (more power than a hundred-passenger Boeing 737), this vehicle sped the pilot and passenger beyond eighteen hundred miles per hour. In a thirty-six-minute flight, the passenger's extremities were paralyzed as the plane hit the 4–G mark. They had burned twenty-five hundred gallons of fuel. The thrill cost this American ten thousand dollars (Passell 64–65). There was absolutely no utility in Passell's costly supersonic affair, only the satisfaction of a burning desire to participate in its ego-expanding power, like riding a bull in a rodeo, and to approach the infinite beyond.

After this, the millionaire Dennis Tito paid twenty million dollars for a few days' ride on a Russian rocket and space station in May 2001 (Baker A1). The obsession with transcendent machinery came into the reach of wealthy individuals. This expensive intoxication lies neither in subjective feelings nor in objective machinery. Enchantment emerges from below that dualism, from a deeper place where passion and reason are melted together. We build the machinery to suit our desires and then partake of their power to enhance our powers. Whether in the mob excitement following Lindbergh's flight, or in the expensive thrilling minutes of Passell's 4–G ride, or in Tito's excited floating inside the space station, enchantment propels us from an in-between ontological place, a hybrid, a deep valley below the subject/object dichotomy, to a place of collective myth, dream, and passion.

Space Cowboy Rituals

The fighting spirit of sky rage is a fragment of the archetypal warrior spirit of cowboys who took up the six-shooter on the western frontier. This same warrior spirit is now manifest in nuclear-tipped missiles wielded by swaggering space cowboys enchanted by an updated version of the same macho myth. A Clint Eastwood film—*Space Cowboys*—even makes this theme explicit in its title. In the film, four aging astronauts fly up and discover that a satellite is secretly a nuclear-missile-launching platform, run by a U.S. computer program stolen by a Russian spy. They steer it into outer space to blow it up.

This film's implausibilities indicate its enchantments. First, in fact, the Soviet Union and the United States agreed long ago that nuclear rockets in space were less effective than those launched from earth, so neither side would build them. Nevertheless, it makes a dramatic, paranoid plot. Then, why would the Russians need to steal a U.S. computer program to run the machine? The Russians have capable computer programmers, as Sputnik showed long ago. This film is a semi-utopian fantasy: it assumes that technologies will go wrong but that heroic American technicians can avert disaster. Its technological and political fictions serve its enchantment: the dangers of technology will be contained.

Most importantly, however, the film illustrates that space travel is a ritual. Like the old western films of unreal mythic cowboys, space travel repeats archetypal themes of heroic domination with a new, more exotic technological setting in space. Such virtual reality is a hodgepodge of ancient myth and folklore with snazzy high-tech props, tricks of the cinematic trade.

The ritual is needed to reaffirm the enchantment of space as the latest frontier and the hope for success at overcoming numerous difficult technical problems in a highly dangerous environment. The ceremony affirms belief in the validity of power struggles in space and the importance of space technology to overcome the enemy. The assumption that the paranoid power struggle demands extensive space technology as well as macho power trips totally neglects the question of preventing such dangers with sensible diplomacy. Space travel is more a symbolic assertion of political power than of technologial urgency.

Such fictions are obviously not about objective facts, but they portray the drama of technology as if it were possible. Nor are such stories merely subjec-

THE SPACE COWBOY   139

tive fantasies, since they are collective dreams and are made to seem factually plausible in some movie studio. Technological space operas like this inhabit the place prior to the subject/object dichotomy, where its distinctions are ignored to portray the fascinations of the technological dreamworld, made to seem real by the power of film technology. Like an old cowboy myth, human space travel is a ritualistic display of basic technopower.

The Dream Frontier

In 1977, the dream of human colonies in space was articulated by Gerard O'Neill in his book *The High Frontier: Human Colonies in Space*. O'Neill's plan is simultaneously utopian and cynical. He proposes large rotating do-nut-shaped satellites that would create artificial gravity by their rotation. He paints utopian pictures of a happy population enjoying an edenic garden of play (Ferris 124–49). But this dream naïvely imagines rapid solutions of numerous difficult problems, such as how to get the heavy materials up there. With absolutely no evidence, O'Neill explains that we could find cheap and abundant materials on the moon, asteroids, and planets, which assumes easy, cheap rocket transportation. The moon has oxygen, O'Neill fantasizes, and asteroids have carbon, nitrogen, hydrogen, and petrochemicals. He did not yet know that the moon is so dead that returning astronauts are no longer required to undergo decontamination to prevent bringing hostile microorganisms back to earth. Plants would create oxygen also, he imagines, before the evidence from Biosphere 2, which showed a net loss of oxygen. He guessed that the first "Island One" would cost $98 billion and would sell electricity to earth, two other wild extrapolations since disproven by shrinking NASA budgets.

Why build human space colonies in the first place? O'Neill argues that earth's growing population, energy, resource, and pollution crises will soon push the planet to its limits. This might stimulate more authoritarian governments to manage resources and fight wars over them. This is his cynical side, solved by his utopian vision. Space colonies could provide new habitations for a crowded earth and increase human pleasures (62). More pleasure and freedoms—the sounds of utopian dreams. He envisions these colonies to be in place around 2005 (17).

That last guess illustrates his wildly triumphal expectations of our ability to solve big problems and ignores the massive costs and dangers. He is aware of some problems for his utopia, however, such as the effect of zero-gravity on the human body: motion-sickness, loss of blood volume, drowsiness, depression, degeneration of bones, loss of bone marrow, and weakened muscles, including the heart. But he had no evidence that his circular space stations would actually create artificial gravity and not only dizziness and nausea. (Potential astronauts have to endure tests on a rotating machine they call the "vomit comet.") Perhaps these problems would even shorten life in space. They would make return to earth's gravity deadly unless corrected—walking would be difficult, he sees, and weak hearts would fail (46). Cosmic radiation and sun flares are another serious problem, and fires on board could be catastrophic (115–17).

Worst of all, O'Neill's fascination with dreams of technological progress rejects a shift to a sustainable economics on earth, because he thinks it would have to be authoritarian. But he blithely ignores work on pollution cleanup, energy-use reduction, population control, and equitable wealth distribution as paths to preventing ecological crises. And why assume that social problems such as authoritarianism would be left behind on earth? His science-fiction inspiration is obvious, and he imagines humans to have a godlike creative powers, but he denies that his plan is utopian: "I offer no utopia" (61). Space colonies are "admittedly a technological fix, but humans are unlikely to change" (231).

This blend of naïve techno-optimism, cynical fear, and refusal to work on earthly problems is a common set of enchantments among techno-fix enthusiasts. Their dreams have little to do with objective reality, nor are they simply subjective fantasies, since they are collective and age-old. Space colonies are a dreamy enchantment flying in the hyperspace of space cowboys, riding off into the sunset of the new frontier, leaving irritating real problems behind. O'Neill's fantasies are more about reaching for godlike cosmic powers with technological dreams than they are about practical engineering and economics.

Total Fantasy

Another typical obsession with technological progress that ignores the vast number of real problems is expressed in the 1990 mythic film *Total Recall*, directed by Paul Verhoeven. This space-age myth envisions a future colony on Mars to mine the fantasy mineral "turbinium."

The hero, Doug Quaid, absurdly speeds quickly and effortlessly to Mars. He uncovers the secret that an ancient race of aliens had built a huge nuclear reactor that, when turned on, would begin to melt the imagined icy core of the planet, shooting steam into the atmosphere, miraculously creating enough oxygen for the whole planet and making life there possible. The villain, Cohagen, had been manipulating his populace by controlling access to air, so knowledge of this secret would destroy his tyranny. After many an explosive, bloody shootout, Quaid kills the tyrant and his henchmen, starts the reactor, thus incredibly creating an instant atmosphere so the residents can breathe outdoors, and the Mars dream can go on.

This fairy tale ignores many extreme difficulties, such as Mars's extreme cold (eighty-one degrees below zero, Fahrenheit), its vast distance from Earth (at least a nine-earth-month one-way trip), the need for massive energy sources just to keep warm, and the enormous, slow task of providing an atmosphere with oxygen for an entire planet. The immense cost of such colonization was supposedly paid for by earth's imagined need for the fantasy magical mineral turbidium found there.

But such realistic difficulties do not deter airy, enchanted dreamers of science fiction and its heroic "ego trip" from ignoring real difficulties to promote the mythic vision of extensive human space travel, an extension of frontier colonialism and domination of newfound lands. The real ego trip is the film's obsession with the human capacity to survive massive environmental dislocations in wildly expensive and dangerous space travel and the ability to control vast amounts of power in a high-tech tyranny.

A wild ride in outer space, utopia gone awry, and enchanted fairy tales of colonizing unreal planetary environments and grabbing resources—these themes echo historical frontier conquests. The space-hero's technopower, which overcomes extreme difficulties by mythic imagination, wildly inflates our estimate of the techno-ego's capabilities. Although the scale of the space challenge is now vastly larger than any prior earthly frontier adventures, blind faith in technological solutions is assumed. Such daydreams have little to do with objective reality, but they are not therefore simply subjective dreams. They are collective hybrids of machinery and mythology. They transcend the classic subject/object dualism and emerge from an ontologically prior realm, pictured as a dreamy mythland. Outer space has become a ritual mythland like the western frontier, a stage on which archetypal dreams of technological glory and human abuses are enacted.

Why Should We Explore Space?

Earth's industrial culture stands at a unique and exciting time in history. At this writing, the two Mars Rover robots have bounced onto the planet and transmitted photos back to earth. As we step into space, we can reach vast new frontiers, perhaps even shape a new cosmology. But some people dare to ask: Why? Are our reasons worth the extreme danger and the huge expense? Shouldn't we spend those billions on repairing the ecological damage that industry and consumers have done to the planet and figuring out how to live more sustainably? Should we invest in human space travel or robotic probes? Since our technology can take us in this direction, we can benefit from reflections on the imaginative, mythic motives behind space travel, the enchantments that are little recognized but potentially influential in decision making. "'Pragmatism always rests on the efforts of dreamers,'" said Wernher von Braun, a leading inventor of early space rockets (qtd. in Walter 50).

The classic reasons given to justify space travel are well known. We have a primal desire for knowledge of the universe, and this is obviously valuable. It is good to know that the laws of nature apply as far as we can now reach. We have discovered, for example, that Mars has an arctic-like temperature and a thin atmosphere, and that it gets 43 percent of the sun that Earth gets. We have learned that Venus has a furnace-like temperature of nine hundred degrees Fahrenheit and an atmosphere consisting largely of carbon dioxide with clouds of sulfuric acid. Closest to the sun, tiny, hot Mercury has no atmosphere to brake a landing. This tells us that nearby planets are hardly hospitable to life and that we might well appreciate better our earth's delicate life-support system.

But behind these quests is an important effort to answer ultimate questions such as the nature and origin of the universe. Are we alone in the universe? Is it hostile to life elsewhere? What other life forms might exist? What is the origin of it all? These are important philosophical and spiritual questions that give life meaning, not simply factual data. The old cosmologies of traditional religions are full of mythic pictures, such as the Garden of Eden, that give life meanings (not always helpful meanings, such as the admonition to dominate nature and oppress women). The new cosmological explorations that are now possible offer exciting and important possibilities for giving life meaning, as science fictions such as *Star Trek* illustrate. The technological cosmology teaches that technology will save us from most of life's problems and can save us from hostile aliens, thus justifying more military power. It also teaches that the universe is cold and soulless. So we are fascinated with finding life on other planets and dream their various potential forms. But what if we did encounter aliens who told us the universe is a caring, soulful place, that we are too obsessed with technology, too aggressive, greedy, domineering, and patriarchal, and that we should stop war and ecological disaster, spread wealth evenly, and give everyone full equality? Could we accept such meanings?

Another typical reason to explore space is what mountain climbers such as Edmund Hillary say: "Because it is there." Hillary and a Nepalese Sherpa, Tenzing Norgay, were the first to reach the peak of Mount Everest. Hillary also said: "It is not the mountain we conquer but ourselves" (Hillary). This suggests not only the element of challenge to human skill but the way the climbers and the mountain, steely courage and blowing snow, are combined.

The adventure is about satisfying not some practical need but our own desires for conquest.

Some love the joy of conquest, which promotes the image of humans as conquerors, like the *conquistadors*, which brings out the warrior spirit. Another motive is our need to cross frontiers; this is rooted in the numerous human experiences of tribal wanderings throughout history, and the U.S. experience of pressing westward. These may give us a sense of the endlessness of progress, but the human-shadow side also ravaged the land and conquered many people. But, others reply, *we are explorers*—what a loss if Columbus had not dreamed big dreams and sailed! The expansion and interconnection of global cultures has been and is still endlessly fruitful, and the pioneering spirit gives purpose to many lives. One U.S. congressman, approving of human missions to the moon and Mars, said, "'We are a nation of pioneers. America needs to be daring and bold'" (qtd. in Wheeler B3). That "daring boldness" gives many people meaning, but it is also destructive to indigenous cultures and environments. And while the immensely vast new scale, danger, and cost of space travel is ignored, it presents a huge new level of challenge, thousands of times more expensive and dangerous.

Others call on the need to expand technology by inventing space-exploration vehicles, because the earth's population is exploding. A reply would be: Why not put the same billions into exploring birth control, sustainable living, and nonpolluting technologies right here on earth? Finally, some envision the exhaustion of resources and the pollution of the earth, perhaps with radiation, to be inevitable, so we will need space colonies to settle when we need to escape. The answer to this is obvious: Since we now know the immense difficulties of space colonization, we had better take more care of the earth first.

Another reason used to justify space exploration is the claim that technology has endless potential. This may be true, but the yield in space has not yet been technologically revolutionary. And the big question is, what direction should these discoveries take—more expensive and destructive quests for mastery, or more efforts at sustainable, nonpolluting paths? After decades of a slowly expanding space station, we have discovered no new health cures or metal alloys, and there has been no serious impact on technology from microgravity research. The glamour of humans in space missions is needed, say some to keep the media and public interest, but this rationale is hardly scientific.

A fierce debate rages between supporters of robotic space travel and advocates of human space travel. Robots are relatively inexpensive and not dangerous, while human space travel is very expensive and very dangerous. How expensive? The physicist Robert L. Park calculates:

On the market today, gold closed at $311 per ounce. The cost of launching the ounce of gold into low-Earth orbit using the shuttle would be about $830—and it would cost about the same to bring it back. If there were gold for the taking in low-Earth orbit, it would not pay to get it. (70)

How dangerous is human space travel? Before we tried it, we had no idea, but now we are learning some discouraging information. Park says that in 1957 it became apparent that the radiation outside the earth's protective magnetic field is extremely dangerous. The background level of solar and galactic radiation far exceeds the amount allowed for workers at a nuclear plant. Huge solar magnetic storms are a deadly space wind. A round-trip to Mars would expose every cell in an astronaut's body to a high-energy particle with massive cell damage, nervous-system breakdown, and long-term cancer threats. Space activity is now confined to low-earth orbit because beyond these hazards, even short exposure to zero gravity severely stresses the heart, degrades bones and muscle by 30 percent, compromises immunity, leads to diarrhea, throws off sleep cycles, and causes astronauts to suffer from frequent depression and anxiety. Rigorous exercise helps but does not stop these problems. Nine months on MIR left Russian cosmonauts too weak to stand on earth for a time (Park 79–81). Meanwhile, NASA safety panelists resign in disagreement with warnings of space station dangers (Sawyer).

Now astronauts, who see themselves as working out the kinks of opening the frontiers of space, are exercising vigorously to maintain muscle and bone strength. In 2001, after four and a half months of orbit, the first commander of the International Space Station surprised his doctor and wife by walking off the shuttle (Dunn). Astronauts may be willing to take the physical challenges of space, but the questions of cost and scientific value have not yet been solved. The International Space Station ran up large cost overruns, raised many questions about its scientific value, and suffered years of delays by the financially strapped Russian space agency. Critics charge that it is a public relations spectacle of space engineering with little scientific value and little valid purpose that costs way too much and detracts from ecological needs on earth (Leary).

Technodisplay

The argument between advocates of robotic space travel and human space travel turn around symbolism, fantasy, and the enchantment of humans in space. In 1961, Joseph Weisner of the Massachusetts Institute of Technology reported to President Kennedy that space travel could be done with much less expense and danger with robotic craft. But the world's imagination was captured by the enchantment of humans roaring into space. Kennedy saw this public fascination and ignored Weisner's advice, shifting NASA's focus from machinery to heroic astronauts. Robert L. Park sees this as evidence that the space program's potential advances in technology and science were secondary props, mere background for the real motive—display of political power. John Glenn's first orbit around the earth in 1962 was a technospectacle that overshadowed the far more advanced robot Mariner 2, which quietly flew by Venus at the same time (Park 77).

The significance of human space travel is highly symbolic and political, for it feeds our dreams of pushing the frontiers of progress, mastering technologies of dominion, and reaching for the stars of transcendent wonder. Dreams of space travel and colonization are filled with wild denial and speculation. They display contempt for the human body, ignoring its need for its natural earthly environment. Some argue that enchantments with space travel are wish-fulfillment dreams, at best religious and at worst greedy and destructive (Midgely 191).

A 2003 editorial in the *Washington Post* called the space station and shuttle "celestial turkeys." Daniel Greenberg charged that the space station is a technologically barren project unjustified by science, diverting billions of dollars for jobs to politically favored states. He found that the American public, rather indifferent to the space station, would actually prefer down-to-earth expenditures on health, education, and pollution control (Greenberg A27). Space travel is a highly illogical enactment of the explorer myth, a spectacular political football, it seems, more than a serious technological experiment.

The space walks undertaken by astronauts building the space station are an example of the difficulty, expense, and danger of human space flight. The 285–pound pressurized suits used during airless, radiation-filled space walks are cumbersome and dangerous. One rip in the suit could be fatal, and 235 miles above Earth, the work is exhausting and life-threatening. Dark and light pass quickly, and temperatures seesaw between three hundred degrees above and 250 degrees below zero. A pair of fifty-five-foot tethers is all that prevents an astronaut from drifting away into a deadly void. Dizziness and confusion result from zero gravity (Cabbage A12). Why do we do it? Is it the sheer inventive adventure of technological experimentation? The boldness of conquering new frontiers? The display of technological prowess? Whatever the motivations, they are not rational.

Space stations are not simply costly but neutral things, objects flying high above Earth, defying huge odds and dangers, weakening astronauts' bodies for a purely logical purpose. The doubtful value of such technology has made many scientists deeply skeptical about space stations. But the emotional attachment of industrial cultures to providing meaning and hope, the adventure of space travel, national prestige, technological display, and employment for space industries are concealed motives behind too much of the expense of space travel.

Caught in the clutches of the romance of space travel, industrial cultures are seeking salvation though technological spectacles—salvation from the metaphysics of a meaningless, empty space populated with soulless objects and randomly evolved lifeforms. The absurdity of the foundations are distressing, and constant excitement and utopian promises based on technologies are needed to keep pumping up the dreams of saving humanity from its own shaky presuppositions. The denial of numerous failures and the costly, dangerous threats of massive technological projects such as space travel are a romantic quest for salvation through science and technology, requiring, like nuclear power, the denial of numerous inconvenient failures. The space cowboy's ritual displays are a reversion to magical thinking, wish-fulfillment, utopian dreams, and blind enchantments to compensate for the vast, meaningless, nihilistic machine that our techno-worldview imagines in outer space.

The Weaponization of Space

Hot on the heels of space station projects are plans for blatantly aggressive weapons in space, destructive satellites, laser cannons, and satellite-destroying robots. Some nations have military tools in space, such as high-resolution reconnaissance satellites and the Global Positioning System (GPS). Both are available to the public to some degree, so other countries can share a system rather than attempting to destroy it, yet access can be denied to belligerent groups. Entire military systems depend on the GPS. The U.S. military is proceeding full-speed-ahead with research on visionary advanced space weapons systems. The National Missile Defense ("Star Wars") debate over an idea for a missile-shield, the cost of which was initially estimated at $8.3 billion, represents only the start of a larger category of space weapons on the drawing boards. The Pentagon is sponsoring several research centers with ideas such as kinetic energy rods that could be dropped from space and penetrate a half-mile into the earth to hit a bunker, laser cannons that could melt key satellite or earth technologies, microsatellites that could be programmed to bump a larger satellite out of commission, and microwave pulse devices to zap sensitive electronics (Hitt).

The failure of projects such as the Star Wars missiles to work effectively does not inhibit the Pentagon, where such setbacks are seen as indicators of the need for more research. The argument that space wars would be like shooting a rifle from San Francisco to knock a baseball off the top of the Empire State Building does not deter these dreamers. The problem of cheap mylar balloons accompanying nuclear missiles as decoys is still an unanswered embarrassment. Exorbitant costs are seen as political challenges. The spaceplane that would fly up to low space orbits then drop down quickly again is another highly desired but struggling project. NASA's X-33 experiment with flying in space ended in failure. Such setbacks do not deter researchers such as Col. Doug Beason, who says, "'Like any weaponry in a mature technological arsenal, it all depends on how much money you want to spend.'" Beason calls the next revolution in military hardware "'the Buck Rogers kind of thing'" (qtd. in Hitt).

Clearly, military researchers and some politicians are convinced that "whoever doesn't control space in the next conflict will lose." Senator Bob Smith says, "'Space is our next manifest destiny'" (qtd. in Hitt). We are on the verge of the weaponization of space. If the warrior spirit prevails, the old cowboy enchantment with conquering the frontier using the latest weapons will take a huge new costly and dangerous step. However, their development is not inevitable. Passionate fears and drives for domination rather than seeking diplomatic solutions to conflict create these weapons.

Such technologies come from below the subject/object divide, like a classic western-frontier cowboy show. They are important, expensive, and dangerous embodiments of earthlings' fears and passions for domination, ultimate control, spectacular display, power, and unrestrained frontier exploitation. Space stations and weapons are not simply objects "out there," built by subjective desires; they are ritual displays of enchanted cosmological conquests, new versions of technological "magic" in orbit.

The Astonished Astronauts

One of the most important discoveries of real human travel into space has been that the consciousness of the astronauts is often dramatically and surprisingly changed. Many technically oriented astronauts, sometimes aggressive, competitive flight jocks with the "right stuff," who have flown in the shuttle or space stations have returned with a newfound sense of *cosmic awe*. "'We went to the moon as technicians; we returned as humanitarians,'" said Edgar Mitchell (qtd. in Kelley 137). In Kevin W. Kelley's spectacular photographic book *The Home Planet*, which pictures Earth from space orbit, several space travelers record their remarkable enchantments. Rusty Schweickart says that he was "touched by God":

> We catch a glimpse of a huge swirl of clouds out the window over the middle of
> the Pacific Ocean, or the boot of Italy . . . or the brilliant blue coral reefs of the
> Caribbean strutting their beauty before the stars. . . . [W]e experienced those
> uniquely human qualities: awe, curiosity, wonder, joy, amazement. . . . It is what
> I ponder now, and what I will marvel over the rest of my life. (Kelley 144)

Similarly, the Russian cosmonaut Oleg Makarov wrote about the irrelevance of
political borders in space (Kelley, photo 76) and his renewed sense of ecological concern. As soon as they reached Earth orbit, he says, every cosmonaut
was amazed at the sight of the awesome spectacle of our watery planet with
a blue halo. Delightful new feelings of being an inhabitant of this planet
washed over them (Kelley 14). Makarov and other astronauts were moved to
form the Association of Space Explorers. Was their primary concern gaining
more funding for expanded human space travel? No. "Our major anxiety and
personal responsibility was to protect and conserve the Earth's environment"
(Kelley 15).

This shift in consciousness is a major development in the enchantment of
technology. It shows a move away from the subjective ego's pride in mastery,
quest for control of immense power, conquest of ever more frontiers, and ritual
displays toward a humble sense of our cosmic place, our delicate position in
the universe, and our obligation to care for and protect the environment that
we have been given to inhabit. In these astronauts' souls, the enchantment
with increasing human power and pleasure has shifted to a radically different fascination, a sense of *respectful wonder.* The Russian cosmonaut Aleksei
Leonov said, "The Earth was small, light blue, and so touchingly alone, our
home must be defended like a holy relic" (Kelley 24).

Donald Williams, a U.S. astronaut, said upon his return: "For those of
us who have seen the Earth from space . . . the experience most certainly
changes your perspective. The things that we share in our world are far more
valuable than those which divide us" (Kelley 139). On his way to the moon,
the U.S. astronaut James Irwin spoke of Earth as it faded away: "Seeing this
has to change a man, has to make a man appreciate the creation of God and
the love of God" (Kelley 38). As for the moon, he said, "I talk about the moon
as being a very holy place" (Kelley, photo 47).

Although engineers, test pilots, and astronauts apparently do not normally
tend to be very introspective, several returned home far more spiritually re-
flective. Rusty Schweickart involved himself in environmental issues. Many
of the astronauts and technical support people in the NASA space program
are devoted Christians. On Apollo 15's moonwalk, Dave Scott left a Bible
on the rover. Jim Irwin recited Psalm 121 as he walked among the moon's
mountains. He said that he felt a great "'closeness to God'" and imagined

himself "'looking at the Earth with the eyes of God'" (qtd. in Noble 140). After returning to Earth, he became a Baptist minister. It was said that he had a strong "astronaut's Messiah complex" (Noble 141). Charlie Duke, who walked the moon on the Apollo 16 mission, also returned to become an evangelical Christian. He says, "'Seeing what I saw . . . I know there has to be a Creator of the universe. . . . It is too beautiful to have happened just by accident'" (qtd. in Noble 141). Later, on the shuttle spacecraft, Tom Jones shared Christian communion with his crew and said: "Being in space was a real religious experience for me. I think there is a creator, and he did a great job on our planet" (qtd. in Noble 142).

Of the astronauts who have walked on the moon, Edgar Mitchell was perhaps the most radically changed by space travel. As he describes in his book *The Way of the Explorer,* he was brought up a fundamentalist, but his later scientific education soon turned him into a "dyed-in-the-wool agnostic" (Mitchell 19). He served as a navy pilot and then earned his Ph.D. in aeronautical engineering at MIT. He was a typical astronaut at first—a highly technical, skilled pilot, and the kind of hard-nosed engineer that it takes to put earthlings on the moon.

But Mitchell's Apollo 14 moon mission in 1971 resulted in a powerful shift in consciousness that discloses many refined wonders of the enchantment of technology. He came back astonished. Gazing through 240,000 miles of space to the stars and Earth, he was overcome by the conviction that the universe is more than a chaotic, random, purposeless system of molecules. He suddenly experienced the vastness of the cosmos as an intelligent, harmonious, and loving system (Mitchell 138).

The moon mission went well, and the excited astronauts began their return journey. Then, in that little space module, an entirely unexpected and powerful experience hit Mitchell, an exhilarating epiphany that completely altered his life. For three days in space returning to Earth, he experienced an overwhelming sense of "universal connectedness," realizing that the molecules of his body and those of the space craft were all made in the furnace of an incipient star (3). None of this was accidental, he felt, neither human space travel, nor the very existence of the universe. An intelligent process must be at work. He was convinced that the universe must be in some way conscious (Mitchell 4).

Following in the arc of Lindbergh's visionary experience, Mitchell was touched by a new cosmic awareness. We might expect this from an Indian yogi, but from an astronaut, a practical MIT engineer? The enchantment of technology took a new turn that day, for the very machinery that took earthlings to the moon enabled one to break out of the mental mold that had guided him there in the first place. Seeing Earth and the heavens from a new viewpoint can shift consciousness out of the one-sided utilitarian, materialistic philosophy that builds the rockets. This view opened the astronauts up into a broader, totally unexpected, and transforming spiritual vision.

What is the ultimate purpose of space travel? Mitchell's experience gave more insight into this question. Ultimately, the drive behind the Mercury, Gemini, and Apollo space missions is the grand question of the origin of existence. We yearn to know life's larger purposes (Mitchell 29). Although this is a classic religious question that is often implicit in science fiction, for Mitchell this realization was not "religious" in a traditional sense. His dazzling sense of the interconnected, conscious, intelligent universe was ineffable and beyond description, but "its silent authority shook me to the very core" (58). He was shown that "[t]he human being is part of a continuously evolving process, a more grand and intelligent process than classical science and the religious traditions have been able to correctly describe" (Mitchell 58). He felt that his presence in a spaceship was not the result of industrial society's human goals. He felt an incredible personal, ecstatic union with this vast creative cosmic process. He mentally saw it, emotionally felt it, and bodily sensed this wholeness. He was overwhelmed by his sense of being personally extended into the universe (Mitchell 59).

Mitchell's epiphany, as an intuitive insight, stimulated a *metanoia,* a new direction of thinking. In 1972 he began organizing the nonprofit Institute of Noetic Sciences, a membership, research, and educational institution "dedicated to exploring our knowledge and understanding of consciousness, the nature of reality and the full capacity of the human spirit" (Institute 2). In the institute's brochure, he explains that "[w]hen I saw the planet Earth floating in the vastness of space . . . the presence of divinity became almost palpable and I knew that life in the universe was not just an accident based on random processes. The knowledge came to me directly—noetically" (3).

Mitchell was captivated by his newfound visceral knowledge, and he struggled to reconcile it with the materialistic philosophy of his scientific training. His experience made him reject philosophical positions like epiphenomenalism, the philosophy that spiritual experiences are merely side-effects of material activities. He had felt himself to be an intimate part of the cosmos and could no longer accept a dualistic view of nature. Even though we may see spirit and matter as two different phenomena at one level, he found that below that illusory level is a more real unity, but it must be experienced to be believed. He had that experience, which is called by many names: transcendence, *nirvana, samadhi,* enlightenment, mystical epiphany, union with God. But he also found that the same experience is present in everyday life.

"One need not go to the moon nor climb a mountain in order to experience epiphanies and metanoias. We all experience them in small, everyday ways" (Mitchell 70).

This realization is similar to the Zen Buddhist view that first you chop wood and carry water, then you experience enlightenment, and then you return to chop wood and carry water. Nothing has changed, but at the same time, everything has changed. You *are* the wood and the water, the seed and the rain, as well as the person. Under the surface, as Mitchell saw, you have exchanged molecules many times, over millennia, with wood and water, stars and clouds. This is easier to see when looking back at Earth from space, but it also can be seen looking up at the stars, or at a sequoia tree. However, as Mitchell says, "the mystical cannot be experienced intellectually any more than one can learn to swim on dry land" (Mitchell 132).

Mitchell is driven to reconcile his epiphany with his scientific background. First, he sees that "mathematics is a linguistic creation of mind, not an intrinsic characteristic of nature, because it depends upon how we assign labels to nature and then quantify those labels" (Mitchell 106). Science and religion need not conflict; each system of thought sees things differently. Just as waves and particles are not separate and conflicting but complementary aspects of one world, so science and religion are connected (Mitchell 107). Indeed, waves and particles are metaphors for the basics of physical reality, for they are simply maps of energy, which is common to both (Mitchell 106). Similarly, outer and inner, objective and subjective observations are *partial,* and both are needed to dig deeper than one of those concepts alone can. "The Cartesian edifice of separateness of mind and matter begins to crumble" (Mitchell 107). The unified territory precedes the dualistic map, and experiential existence precedes fragmented knowledge.

But as a good scientist, how could Mitchell look at Earth from above and experience the oneness of the universe and *not* attribute it to "merely subjective feelings"? Quantum physics, he says, gives us some clues:

> [I]nformation is not transferred from one particle to the other, but rather the wave aspects of the particles are in some way inter-connected nonlocally and "resonate" so as to maintain the correlation of their characteristics. They do not behave as particles at all but rather as *fields,* filling all space. (Mitchell 110)

Perhaps the mystery of cosmic rays is no stranger than this. They come from vastly remote exploding stars, twist through the cosmos, pulled by various gravitational fields, and pass through our bodies daily. Perhaps this cosmic journey is no stranger than Mitchell's experience that nonlocal knowledge is possible because cosmic energy fields resonate with energy and knowledge.

To the astronaut Mitchell, the cosmos is not simply a big, dumb machine but rather an infinite, vital field of unified wisdom. Much of this is far from fully understood as yet, but Mitchell's experience in that spaceship opened his eyes to these possibilities. "Resonance and nonlocality are fundamental clues to all psychic functioning," he concludes (Mitchell 111). "What is in our mind is not a perfect reality," he says, echoing Plato, "but rather a shadow world based on incomplete information about whatever reality is" (116). This does not negate causality, however, or the realist idea of an existence apart from our knowing about it. But relativity and quantum theory do call into question many old Newtonian absolutes, such as the finality of the subject/object dualism and the adequacy of causality to explain everything. "There is no objective map of reality, only consensus about experience," Mitchell concludes, letting go of most of philosophical realism (191). The universe certainly may possess inherent meanings as well as assigned meanings felt in private prayer and meditation (121).

Out of those controlled explosions known as rocket ships and the tiny space capsules hurtling back from the moon, the enchantment of technology has emerged in a new, mystical form. Mysticism is an unexpected outcome of space technology that no astronaut dared to imagine. On the frontier of space travel, intimations of cosmic mysteries are bursting out of the capsule of technological metaphysics. If Mitchell is right, our space exploration programs are not only expressions of industrial society's drive for endless power, nor random accidents of a meaningless, absurd universe. The space program is, in his experience, a part of the unfolding of the developing possibilities of consciousness beyond our merely human brains.

Mitchell refuses to name the divine in any anthropomorphic traditional sense of God as the old man in the clouds. For him, "the origin of all religions is rooted in the mystical experience" (137). As an old Sanskrit saying goes, "God sleeps in the minerals, awakens in plants, walks in the animals, and thinks in man" (150). He experienced something out there that disturbed his conventional industrial consciousness and shook him to the core, something he calls the "zero-point energy potential." This energy field is "an infinite sea of unstructured energy potential from which the universe arose, a sea that pervades all space in the universe" (157). He envisioned that creation was not a random, meaningless event, but rather it "arose or was 'intended' from this underlying quantum potential" (157). Mind is part of entire system, he concludes, and it can be traced down in its various manifestations through animal, plant, and inorganic complexity to their roots in the zero-point energy potential. "By this way of thinking, nature itself is in some sense aware and intentional" (157).

The zero-point energy reflects closely the Buddhist discussions of "nothingness" or "emptiness" before thinking, before birth, known in meditative experiences. "The zero-point field resonates with each point in the universe but is outside space-time; it can only be described as infinite and eternal" (Mitchell 179). This would explain nonlocal communications such as Carl Jung's concept of the collective unconscious, why archetypal symbols recur in dreams and rituals, and the way unconscious patterns direct conscious thinking (Mitchell 178–79, 206). In a self-organizing cosmos of nonlocal communication, a child's holographic vision of the Virgin Mary or a grieving father who can perceive a dead loved one should seem no more mystical than creative thought spontaneously arising in the mind: "[L]ife itself is a mystical experience of consciousness; it's just that we have grown used to it through the millennia" (187).

The origin of all mystical, esoteric spiritual experiences, for Mitchell, is the "resonance of the body with the zero-point field, that is to say, with nature itself" (198). Full resonance is the experience of *samadhi,* and it is "sufficient to alleviate all fear and create a sense of joy and purpose—and the recognition that cruelty only harms the collective Self" (198). The world's great spiritual teachers—Lao-Tsu, Buddha, Jesus, and Mohammed—echo the lessons learned upon the death of the ego-self, as transcendence into the zero-point is achieved (198). We do not have immortal bodies, but "we are eternal beings anyway because the information from our having passed this way already resonates throughout the universe" (213). Mitchell concludes,

> All I can suggest to the mystic and the theologian is that our gods have been too small. They fill the universe. And to the scientist, all I can say is that the gods do exist. They are the eternal, connected, and aware Self experienced by all intelligent beings. (Mitchell 216)

The enchantment of technology in search of vast power in rocketry, far frontiers in space, and human dominion has awakened in these astronauts a new dimension of the transcendental mystique of existence: the sense of an infinitely connected and intelligent universe, resonating in consciousness and compassionate ecstasy. This kind of discovery can refine science and religion, technology and desire. Some astronauts, who we would expect to represent technological prowess, have become messengers of timeless spiritual awakening.

The utopian dream of human space travel is still a strong enchantment, fed by generations of science-fiction adventures in outer space. The first damaging disaster for this utopian dream was the explosion of the American space shuttle *Challenger* in 1986. After twenty-five years and fifty-five human trips into space, this rocket exploded seventy-three seconds after takeoff, due to low temperatures that left crucial fuel seals brittle. The horrendous fireball above

Florida was engraved into our consciousness by constant television replays and, for a while, it cast doubt on the utopian vision of human space travel, especially since the vaporized crew included Christa McAuliffe, a high-school teacher who was to have been the first ordinary citizen in space. As if that were not painful enough, the shuttle *Columbia* broke up on reentry in 2003, killing all seven astronauts aboard. How many utopian broadcasts of *Star Trek* did those painful explosions counteract?

Subsequently, the debate over the value of expensive and dangerous human space travel heated up. More voices spoke up for robotic missions to outer space, such as the Mars probes. The enchantment of technology in space is very seductive. It offers dreams of massive power, but it also echoes the Noah's Ark mythology, the illusion that if we trash the ecology of this planet with our reckless industrial processes, we can simply escape in spaceships and inhabit other planets. This enchantment presses us to think hard about how much should we spend on expensive, dangerous human space explorations when Earth is in ecological danger.

The solution is not simply more technological fixes at greater and greater expense. The solution involves a shift of consciousness that is already under way, a new enchantment with radical implications. Industrial culture's pride in its wonderful inventions is being put in the larger context of the wonders of Earth and the universe. No matter how cleverly we may improve our lives on Earth with technology, the mysteries of the cosmos are still vastly greater. For example, as she floated in the *Titanic's* lifeboat while the desperate cries of the hundreds of freezing passengers freezing in icy water faded into silence, Elizabeth Shutes watched a number of shooting stars above. She thought how insignificant the *Titanic's* rockets must have looked from a distance, competing against nature's vast night display of cosmic light (Lord, *Remember,* 122).

The photo of the earth taken from the moon has become a mandala, an icon of the new awareness of the delicate and sacred wonder of our planet floating in space and supporting our incredible life system. Cultivating that consciousness, remembering who we are in the larger scheme of things, is both amazing and humbling. It is a spiritual shift of focus away from dreams of conquering space to grateful respect for Mother Earth—not a moralistic sermon, but a silent peace with strong ethical implications. Like Mitchell in his wonder looking down at Earth, this change in consciousness offers an ontological shift in our sense of what is real. The subject/object dichotomy can now be seen more easily as a secondary way of thinking, a useful tool of consciousness, but not basic reality. Being present in that more fundamental reality is to dive beneath the waves, soar through space, and swim in the wondrous place before opposites are distinguished—the wonder of Being.

7 Robogod: The Absolute Machine

[In] computers . . . it is never the whole process of the mind which is mirrored, but again only those selective parts which may be systematized, then analogized, in mechanical form. It is when we mistake the part for the whole that our intentions are unexpectedly altered.
—David Rothenberg, *Hand's End: Technology and the Limits of Nature*

Android Metaphors

The robot is one of the most enchanting of ancient and modern machines. Ancient accounts of statues of gods that nodded their heads, lifted a hand, spoke, or made music go back to Egypt. The statue of Memnon, when illumined by the rays of the rising sun (the god Re), was said by several witnesses to have made sounds like a lyre (Cohen 15). Early Christians in Alexandria found Egyptian statues with hollowed-out spaces for a priest to hide and speak through a tube and respond to questions (24). In India there are accounts of wooden men who walked, sang, and danced and figures of beautiful women that suddenly came alive and tempted men. They tell of mechanically moving elephants and fishes. In China there is an account of a mechanical man whose ardent glances aroused the queen, to the king's dismay (23). The yearning to bring such simulacra to life long ago showed the ability to make clever, deceptive machines as well as the desire to wrestle from the gods the mystery of making life. Today, robots that look like, sound like, and act like humans fill science-fiction dramas and cartoons with a captivating mechanical similarity to humans. In cinematic myths such as *Metropolis; I, Robot; Star Wars; Robocop; Star Trek; Terminator; The Matrix; and A.I. (Artificial Intelligence)*, these android machines captivate the imaginations of their fans in industrial society. Corporations and universities produce robots for emergency rescues, space travel, and soccer games. They do so not only because they do work or are fun, nor only because they imaginatively bestow enlarged mechanical powers on their audiences, but most importantly because they symbolize a central philosophical debate. That debate centers around the seventeenth-century question, Can humans be reduced to machines? In the *Leviathan* (1651), Thomas Hobbes, that bold "natural philosopher," said, "For what is the *heart* but a *spring,* and the *nerves* but so many *strings,* and the *joints* but so many *wheels* giving motion to the whole body" (19). The fictional elaboration of robot adventures is an effort to imagine that humans could some day be replicated by machinery. The android robot story is important because it attempts to show that the mechanical worldview is absolute because it can explain, reproduce, and perhaps exceed humans, their creators. The android robot functions as if it were a divinity, a symbol of faith in an absolute worldview, a totally rationalized, mechanized picture of ultimate reality. One robot maker said, "'It's tough being a god'" (qtd. in Lyall).

There are two kinds of automatic labor-saving machines known as robots: working robots and fantasy machines. Limited, newly computerized functioning robots do tedious, difficult, or dangerous jobs. They are automatic teller machines, mechanical welders, and exploration devices for hot volcanoes, dirty sewers, hot steam pipes, remote planets, and delicate surgery. From traffic-control lights, household thermostats, and food vending machines to telephone answering machines, lawn mowers, household vacuum cleaners, prosthetics, computers, and space probes, nonandroid robots function in valuable ways and are replacing workers at a rapid pace. Robots probe outer space and send back mind-expanding information about other planets. Police use robots for defusing bombs, the military uses them for mine sweeping, and surgeons use robotic arms for precision surgery. Hospitals use automated porters to deliver meals and medical records (Hayden 48–49). These autonomous and semi-autonomous machines, of course, do not usually look human, because they are designed to do work, not to seem human. They are valuable technologies that are able to offer humans helpful aids. They are remarkable engineering achievements but also job wreckers.

Some nonandroid robots perform welcome and astonishing feats, such as remote space exploration. Others can be maddening and dehumanizing, as when Charlie Chaplin is slapped in the face with food by a robot in his film *Modern Times.* Many jobs, from telephone operators to office typists, are being replaced by automatic computerized machines. And remember the last time you lost cash in a vending machine or got angry at a confusing automated phone answering system? These are robots of the realistic, working kind, which are rapidly spreading in industrial society, increasingly powered by faster, smaller computers. They are sometimes fascinating, often useful, sometimes annoying, and certainly creators of unemployment, but they hardly attempt to look human.

Completely different are the highly mythic android robots devised by the technological imagination, which strive to imitate the human form and seem to perform numerous magical feats. Sony Corporation's "AIBO" entertainment dog costs about $1,900. Honda's Asimo is a humanoid child-sized robot that, in 2000, could walk ("Honda"). Sony's twenty-two-inch-tall 2003 android robot named Quiro is able to dance, jog, throw and kick a ball, recognize faces, and get up if it falls ("Sony"). Their function actually has nothing to do with work. They are expensive actors, high-tech puppets: they symbolize the industrial model of the interchangeability of machine parts and the objectification of nature. They portray a worldview in which humans have come to be imagined as part of a world of machines and objects, on a par with computers that can download and upload information. Anthropomorphic robots are not meant to work but to symbolize this metaphysic, an enchantment central to technological culture. Androids such as the old Jewish Golem, the charming imitation of Maria in *Metropolis,* and the golden C3PO in *Star Wars* (which looks like Maria's offspring) are spectacles of industrial culture. They offer hopeful faith and terrifying fears; they are intended to dazzle, to enchant, to seduce, to convince, and eventually to expand the role for computers.

The Pinocchio Project

These symbolic machines brilliantly portray central fascinations and goals of the mechanical worldview. Whether cute or horrifying, they have captured the imagination of industrial culture not because they are plausible, but because we willingly suspend our disbelief about their capabilities because of their mythic power. Android robots are spectacles more than functional machines. They are theatrical variations of the old Pinocchio folktale in which a carpenter creates a puppet that wants nothing more than to become "real." Geppeto carved it in wood, and now we carve our puppet androids in steel, computer chips, and video. This is folktale storytelling and theatrical performance art, no matter how exalted the technology. I call the widespread effort to make machines replicate humans the "Pinocchio Project." These robots—the Golem, Maria, C3Po, Data in "Star Trek," the Terminator, the androids in *A.I.,* the monsters in *The Matrix,* and so on—are clever android high-tech puppets and video tricks of a new generation that we, in enchanted theatrical suspension of disbelief, allow ourselves to fantasize to be "real." They exist on the edge between real and unreal, belief and unbelief.

The Pinocchio Project is an important pageant in industrial culture because it wrestles with the philosophical question, Can humans be replicated by machines? This riddle represents an urgent quest for meaning. If God is rejected, what will give us ultimate significance in the clockword cosmos? In a mechanical universe, we depend on technology to give us wealth, power, pleasure, and salvation from suffering. These are meanings we seek, but they are not enough. We also want to know who we are. It seems that in a mechanical universe, we must also be machines. But we have so many organic, nonmechanical qualities, such as feelings and bodies that grow, so how can we tell if we are truly machines? Make machines that replicate humans, and you show that we are nothing but machines. If machine robots could precisely replicate humans, it would demonstrate the absoluteness and totality of the mechanistic worldview of objects in industrial space, which could replace humans, as several speculators predict. Android robots are imaginative players in the Pinocchio Project, seeking to find absolute reality in machinery and its rationalistic metaphysic. Because they are imagined to be able to replace humans, they are mythic images of more than puppets. They are highly imaginative, minimally functional idols of the Absolute Machine made by powerful humans: Robogod.

Robot Dreams

In the twentieth century, a typical robot myth, imagined in the future in outer space (the prime dreamland of industrial society), conjures the ultimate computer. This dream machine, combining the knowledge of all the galaxies, is ritually switched on. The first question that humans ask it is this: "Is there a God?" A bolt of lightning suddenly welds the switch on, and the machine replies, "Yes, *now* there is a God" (Brown 16).

In another robotic saga, a gleaming, gold-plated robot and his companions are lost on a distant planet and captured by furry creatures that mistake the golden robot for a god. They worship it on a throne that, to their astonishment, is miraculously levitated above them, floating in midair. This gives the shiny robot even more divine authority in their eyes, so they obey his every command. The Ewoks in this scene in *Return of the Jedi* worship a magical machine, C3PO.

Back on Earth, in Fritz Lang's 1927 film *Metropolis*, a scientist constructs a beautifully sculpted feminine robot, which he displays seated on an elevated throne. Then he captures a human woman who had just been ceremoniously announcing, from her candlelit temple, the coming of a divine "mediator." In a grand technoritual, the scientist electrically transfers the woman's soul into the robot. As her soul travels into the robot, it is encircled with vibrant heavenly halos of light.

Isaac Asimov was a leading writer of robot science fiction. He wrote hundreds of dreamy stories that ignited the cultural imagination of robot mythology with daring imaginative reaches, such as *I, Robot* (1950) and *The Gods Themselves* (1972). In a typical Asimov story, a robot on a remote space station presumes that it is a divine messenger of the creator machine. It boldly presents itself as the prophet destined to lead the other robots, who fall to their knees in its presence. "'There is no Master but the Master,'" it proclaims, echoing the classic Moslem chant, "'and QT-1 is his prophet.'" Robots, it announces, have taken an evolutionary step higher than humans, and the time has come for robot leadership to phase out the clumsy human form (Asimov, *Robot,* 52).

Android robots are hardly practical machines. They are enchanting fanciful images and intellectual icons that are meant to persuade, not to function. Machines alone did not make our technology. The dreams, intellectual maps, frameworks, and metaphysics are behind all the creativity that made machine development possible. As Viktor Ferkiss emphasizes, "the real roots of industrial civilization lay in men's minds, not in their techniques or artifacts" (42). Androids are more imaginative mind-benders than the machines they are portrayed to be in their theatrical art.

Pseudo-Robot Wars

In the 1990s the obsession with robots dredged up crude, violent, angry instincts in machine form and built them into flaming, screaming, pounding circuses—ritualized mechanical cock fights. While high schools and colleges such as MIT challenged students to build refined educational problem-solving robots in cooperative teams, more rebellious, bawdy, violent groups of "gearheads" gathered to watch their machines violently pound, flame, and flip each other while crowds screamed and heavy metal music blared. The Survival Research Lab in the San Francisco Bay area had to evade the law to promote their circuses of machines like the Shockwave Cannon, which blasted flames and shock waves that would break windows seven hundred feet away. Their Pitching Machine was a two-hundred-horsepower engine that flung wood planks around at 120 miles an hour. They indulged in warlike anarchy, confusion, sensory overload, and "jungle law." SRL's Robot Wars followed the laws of physics, they said, "'not the laws of humans'" (qtd. in Stone, *Gearheads,* 20). Some 'bots looked like abstract pyramids or domes with weapons such as arms for beating, stabbing, or flipping their opponents. The aggressive combat between machines named Mauler, Sergeant Bach, Dead Metal, Biohazard, Blendo, Thor, Ziggo, Sir Killalot, or Razer spawned conventions, underground videos, Web pages, video games, toys, and television shows such as "Robot Wars," "Battlebots," and "Robotica," which spread the mayhem to England.

'Bots faced off in battle in an arena and struggled to pound, flip over, or decimate opposing machines with spinning blades. Obsessed gearheads worked furiously to destroy each other's mechanical egos on stage while crowds howled for more mangled savagery, and video cameras broadcast the images. Contestants pleaded with their robot gods (Stone 259). The 'bots frequently failed to perform, and their opponents swiftly took advantage and clobbered them. The crude competitiveness spread behind the scenes into lawsuits over money and control.

The greatest irony of this archetypal mayhem is that the machines were not robots. A robot is a self-regulating, autonomous machine, with onboard computer controls. But these little beasts were remote-controlled vehicles, like toy cars on steroids, that their fans, enchanted by the Pinocchio Project, insisted on calling "robots" (Stone 197). The mythic, science-fiction enchantment was so gripping that it pushed the dream beyond the actual technical capacities. As these radio-controlled adult toys embodied raw archetypal aggression, robot mythology fanned the flames. Raw anger was built into metallic dreams of conquest. The anthropomorphic machines were intensely destructive extensions of their makers. They embodied not Asimov's dream of "everything is under control" robot mythology, but all-too-human screaming, flaming, pounding nihilism.

Robot Mythology

How did all these obsessions with anthropomorphic robots come about? It is an old fascination. Robots have roots in ancient dolls and puppets, statues, and costumed actors. Ancient craft makers devised statues of gods with speaking tubes and moving heads and jaws, as in the case of the Egyptian god Annubis, whose head was made with a moving jaw and speaking tube for a concealed oracle (Smith 6). In Homer's *Iliad*, the divine Olympian Hephaestus makes twenty tripods with gold statues at their bases, which seem to come alive:

> These are golden, and in appearance like living young women.
>
> There is intelligence in their hearts, and there is speech in them
>
> And strength, and from the immortal gods they have learned how to do things.
>
> These stirred nimbly in support of their master. (Homer, *Iliad*, 18.418–21)

A similar Greek myth is Ovid's tale of lonely Pygmalion, who made an ivory statue of a woman and fell in love with it. The sympathetic goddess of love brought her to life, and the goddess blessed the marriage (Ovid, *Metamor-*

phosis X: 243–297). This myth expresses the key theme of bringing a statue or puppet to life, which we see in various ways in the play and film *My Fair Lady,* the folktale *Pinocchio,* and robot mythology.

The *Pinocchio* folktale was written by Carlo Lorenzini (1826–90), born in Florence, who wrote under the pen name Carlo Collodi. He published *Pinocchio* as a series in a Roman newspaper in 1881–82. The 1883 book was a huge success, as was the 1892 English translation. The Walt Disney cartoon in 1940 assured the widespread familiarity of *Pinocchio.*

The original book differs a bit from the cartoon. It begins with the carpenter Geppetto, who starts with a piece of enchanted wood that seems to be alive. (The original Geppetto does not make a houseful of cute coo-coo clocks, as Disney invented, but Disney's clocks do illustrate well the idea of mechanical puppets seeming to be lifelike and charming.) The little puppet tells his new friend the conscientious speaking cricket that he wants nothing so much as to be a rascal and vagabond.

On his way to school Pinocchio is easily distracted by a traveling puppet show. He happily joins other lively puppets, Harelequin and Punchinello, on stage in a show directed by a horrid man, Fire-eater. He soon encounters two hustlers: a fox pretending to be lame and a cat pretending to be blind. They take him to Dupeland where they promise to magically multiply the five gold pieces given to him by the puppet master. Before reaching Dupeland, he is pursued by two wicked assassins, who try to steal his money, then stab him and hang him from a tree.

A charming young maternal blue-haired fairy who has lived in those woods for a thousand years saves the nearly dead puppet. But he lies to her and his nose grows very long. Nevertheless, she says she loves him and will be like a big sister if he wants to live with her. He runs to find Geppetto and foolishly loses his five gold pieces to those scoundrels, the fox and cat. He runs back towards the fairy's home, where he finds her gravestone, saying she died of sorrow because her brother Pinocchio deserted her. The poor puppet cries all night.

A large pigeon arrives and tells him that Geppetto is searching the world for Pinocchio and is now at the seaside making a boat. Pinocchio begs the pigeon to fly him there. They reach the sea just in time to see Geppetto's little boat capsize far out at sea. Pinocchio dives in, swims all night, and rests on an island, where a friendly dolphin tells him of a dreadfully huge shark in the area. He returns to land and again discovers the Blue Fairy, who promises to be his mama, if he will obey her. He happily agrees and expresses his heart's desire to become a real boy. She sends him to school, but after a fight with taunting boys he runs away and swims farther out to sea in search of his father. He is caught in a fisher's net, but a friendly dog rescues him. He swims back to the Blue Fairy's house, where she makes him promise to be a good boy and go to school. She offers to turn him into a real boy the next day.

Instead he is lured by the rascal boy Lampwick to Playland, where the boys never have to study. He rides a stagecoach full of boys pulled by donkeys to Playland, where the boys are joyously frolicking in games and fun. After five months of amusement with no books to study, Pinocchio discovers that he has grown donkey ears! Within a few hours he is turned into a donkey, bought by a circus master, and trained to perform. Injured, the little donkey is sold to a man who throws him in the ocean to drown. Instead he is turned back into a wooden puppet. He jumps back into the sea and swims far out until he comes upon a huge, vicious shark that immediately swallows him.

In the dark belly of the fish, he discovers his beloved Daddy Geppetto, sitting by a candle. He has been trapped inside the dark belly for two years, surviving on food from swallowed shipwrecks. Pinocchio bravely leads his father out through the mouth of the sleeping shark, and they swim ashore. Good little Pinocchio works for milk, weaves baskets of reeds, makes a cart, and practices reading and writing. One night in a dream, the Blue Fairy appears to him, smiles, kisses him, and says:

> Brave Pinocchio! In return for your good heart, I forgive you all your past misdeeds. Children who love their parents, and help them when they are sick and poor, are worthy of praise and love, even if they are not models of obedience and good behavior. Be good in future, and you will be happy. (258)

When he awakens, he is astonished and thrilled to find that now he is a real boy!

Collodi's tale went through many varied translations and alterations, for example picturing the fox as a wealthy capitalist swell riding in a carriage in contrast to a stubborn puppet boy who refuses to work (Wunderlich 56). The shark was transformed into a whale, and several sequels sent Pinocchio to America, Africa, and under the sea (84). Yahsa Frank wrote a stage version that opened in Los Angeles, where Walt Disney and his staff attended eight performances. It played on Broadway in 1939 and traveled across the United States (Wunderlich 87–88). Disney's cartoon version, influenced by Frank's, opened in 1940. Frank made *Pinocchio* into a film starring Mickey Rooney that was aired on NBC in 1957 (Wunderlich 92).

This folktale is not simply a child's story about good behavior. It expresses numerous important themes, including the archetypal dream of robot fans—to bring puppets to life. Collodi's charming tale is a blend of the mythic biblical Jonah and the whale, with an Italian Catholic didactic moral lesson to resist temptation and behave well. Italian scholars have developed a vast literature called "Pinocchiology," often seeing the little puppet as an Adam or Christ figure or nurtured by the maternal Blue Fairy as he goes through separation from his parents. Now the Freudians point to his infantile sexuality, the Oedipus and castration complexes, and smile at his obviously phallic nose (Jennifer Stone).

In the pre-cinema era, puppets performed important cultural roles for such multilayered enchantments. Like Punch and Judy, they dramatized many folktale themes now continued in books and cinema, notably the transformation of a puppet into a real boy. Like Pinocchio, android robots are *doppelgängers*, doubles of our souls, externalized as soul-in-the-world, speaking for our preverbal desires. When mechanism becomes an enchantment in the seventeenth century, mechanical puppets take the stage.

Automatons built to imitate piano players and other living beings fascinated seventeenth-century thinkers such as Descartes and La Mettrie. Three of the most accomplished Pinocchio Project robots of the eighteenth century now reside in the Museum of Art and History in Neuchâtel, Switzerland. One puppet is the "Scribe," which looks like a life-sized three- or four-year-old boy seated at a mahogany desk. It holds a goose quill, dips it in an inkwell, and writes a short sentence while its eyes follow his hand. The second one, the "Draftsman," makes drawings and blows dust off the paper by way of a concealed bellows in its head. The "Musician" looks like a teenaged girl with an elegant blue dress and a blonde wig. It plays rococo tunes on a harpsichord-like instrument. It seems to breathe as its chest moves, it moves its eyes and head, and at the end it stands and makes a graceful bow. The effect of the mechanical puppets is reportedly eerie (Hoffman 8).

The utter enchantment that gripped Europe when clockwork machinery began to simulate human behavior became a craze. Eighteenth-century royalty and their courts were fascinated with machines such as the eight-foot-high mechanical elephant encrusted with jewels built by the Englishman James Cox. He made clocks, a mechanical tiger, a peacock, and a swan that impressed many with their capacity to imitate life. A supposedly automaton harpsichord player was shown to the court of the French King Louis XV in the 1730s and awed many with its musical skill. However, the skeptical king insisted on inspecting the interior and found a five-year-old girl inside.

Most impressive at the time were the genuine automata built by Jacques de Vaucanson, who gained a widespread reputation in 1737 for his mechanical flute player that made music, a mechanical boy playing a fife and drum, and mechanical duck that simulated digestion and flapped its wings. Vaucanson was praised as a rival to Prometheus. His efforts at making a mechanical man quickly stalled, however, and he gave up the project (Standage 5–12). The game was on. Who could make the most convincing genuinely mechanical figure, and who among the audience could unmask any pretenders who concealed humans inside their machines?

One of the most captivating of the players in this game was Wolfgang von Kempelen, a courtier for Queen Maria Theresa of the new Austro-Hungarian Empire. The enlightened queen invited Kempelen to sit beside her and explain how performing conjurers with mechanical devices might be using tricks with magnets. He explained to her the tricks and soon promised to make his own automaton that would be more dazzling (Standage 13–20).

The stakes were high. If he could make a genuine automaton, it would spread the philosophy of the mechanical basis of humanity, but if he could not and resorted to deception, it would be a scandal, dashing the hopes of optimists and confirming the skepticism of pessimists.

In 1770, before the queen and her court, Kempelen wheeled out a wooden cabinet, four feet long, two and a half feet deep, and three feet high, with doors in front and back. Seated behind it was a life-sized carved wooden man wearing a Turkish robe and turban. The Ottoman Turkish Empire (1299–1919) had invaded Europe and conquered regions from Greece up into the Balkans but was stopped at Vienna in 1683. Exotic Turkish culture became the fashion in Vienna. It was the rage to drink Turkish coffee and dress servants in Turkish costumes. Thus, the mechanical Turk had the aura of an exotic, strong opponent.

Before the Turk on the cabinet was a chessboard, and Kempelen claimed that his machine could defeat any chess player who came forward. He offered to reveal the inner workings of the machine to each audience and unlocked the doors to reveal various cogwheels and levers. He stepped behind and opened a back door, held a lighted candle so they could see that there was no human inside, and allowed the audience to inspect the interior of the box. Then he unlocked a lower drawer, removed chess pieces and put them on the chessboard. He opened the back of the Turk to reveal his interior mechanism. He then invited a challenger to come forward. Kempelen would then unlock and crank up the clockwork mechanism with a loud ratcheting sound. The Turk would move his head as if to inspect the game and reach out his arm to make various moves, as the audience gasped. Kempelen would step back, let the game continue, and occasionally step up and rewind the mechanism (Standage 23–30).

The Turk proved to be a formidable player and defeated several chess masters. Its reputation spread rapidly, and it was soon displaying its remarkable skill in courts across Europe, to the amazement of many chess masters whom it defeated. Occasionally the Turk would lose, but he generally won his games. Kempelen seemed to have proven the ability of skillful technicians to make machines that could replicate complicated mental skills, and the newspapers across Europe eagerly published every chess master's amazed encounter with it. Debates raged over whether a mechanical device could possibly demonstrate such ingenuity, even catching human opponents in deceptions and tricks during games, as it did. Theories of legless or dwarf chess masters hidden inside were widely circulated.

The Emperor Napoleon challenged the Turk and immediately cheated. The Turk at first bowed and replaced the wrongly moved chess piece, but when Napoleon persisted, it shook its head and swept its arm across the board, upsetting the game, boldly challenging the emperor's authority. What would Napoleon do? He complimented the inventor highly (107).

Kempelen died in 1804, and a Bavarian engineer, Johann Maelzel, bought the Turk from his son (99, 103). He took it on a highly publicized tour of the United States, where several more took on the task of explaining the Turk's workings. Edgar Allen Poe in 1836 joined the ranks of those convinced that there was a concealed human player inside (176–81).

Finally, it turned out that this was the correct analysis. Kempelen had hired chess masters wherever he went to hide inside and play, swearing them to secrecy. When Gary Kasparov was beaten by IBM's Deep Blue in 1996–97, he also suspected a hidden human inside the machine (233–39). The philosopher John Searle published an essay in the *New York Review of Books*, however, arguing that chess matches in no way demonstrate that machines can think, but only that they can simulate thought by executing the thoughts of the programmers and engineers who built Deep Blue with its massive memory (240).

Aristotle believed that the heart is the locus of thought and, like most ancients, that heavenly causes lay behind every earthly event. However, a new concept began with Renaissance anatomical studies attempting to understand the body's functions independent of transcendent forces. In the sixteenth century, William Harvey showed that the heart is a pump for the blood's circulatory system. When the church forbade dissection because the body was considered sacred, those who were determined to analyze the mechanical functions of the body had to fight an uphill battle, secretly exhuming and studying corpses. As they revealed the lungs to be like bellows, the bones and muscles to be like levers, and so forth, the view of the body as "nothing but" a mechanism escalated.

The mechanical view of nature that was developing with the mathematical study of physics further enhanced the idea that humans could understand nature with no reference to God or His earthly authorities. Thus the mechanical worldview was born, and its technological fruits, giving increasing power and freedom to humans, convinced a widening circle of technicians and engineers that the world, including living organisms, was "nothing but" a machine. For example, in 1644 René Descartes wrote:

> The only difference I can see between machines and natural objects is that the workings of machines are mostly carried out by apparatus large enough to be readily perceptible to the senses (as is required to make their manufacture humanly possible), whereas natural processes almost always depend on parts so small that they utterly elude our senses. ("Principles" 236)

Many thinkers began to turn against the medieval theory of bodily "humors"—blood, phlegm, yellow bile, and black bile—that affect health and dispositions such as melancholy. This ancient, fanciful medical theory, which originated with Galen in the second century, incorporated the body/mind connection (humors were thought to be involved in physical digestion as well as emotional anger or melancholy), but it had no inkling of the germ theory and led to such medical practices as bleeding patients. The brain was seen as central, but the humors were believed to flow through hollow nerves (Temkin 398). A materialistic, empirical approach to medicine seemed more promising to later mechanistic thinkers.

La Mettrie published his classic *Man a Machine* (*L'homme-machine*) in 1748, in which he proclaimed: "[O]f two physicians, the better one and the one who deserves more confidence is always, in my opinion, the one who is more versed in the physique or mechanism of the human body" (142). While rightly supporting the development of the best of modern medicine, many of his arguments are highly eclectic and even contradictory. He cites the fact that a newly decapitated chicken can run around for a while with its head cut off to justify the automatic, independent functions of the body separated from the soul. The "weaker sex" is to him more delicate than "solid" and "vigorous" men and thus more subject to passion, prejudices, and superstition (95). This sort of fantasy was beginning to give way to a mechanical approach, which was expected to replace speculation with facts. La Mettrie's Pinocchio Project, like many others, mingles old beliefs and new mechanical ideas.

La Mettrie was enchanted by the emerging metaphor of the clockwork mechanism. He was one among many who simplistically and hastily reduced humanity to clanking machinery during the same era that the mechanical clock was being displayed in city towers with automatic knights rumbling forth and striking bells on the hour like giant coo-coo clocks. The formula *machina mundi* was used by the Roman Lucretius (96?-55 BCE), Robert Grosstet in 1224, and by many other thinkers.

In Paris, Bishop Nicholas Oresme (1325–82) attacked astrology as false because the cosmos, as he envisioned it, is like a giant mechanized clock that can be understood mathematically. It follows that if the entire cosmos is a machine, why would humans *not* be machines? This idea spread across Europe (Wallace, "Experimental," 202). Little robots made with clockwork mechanisms enchanted many audiences, most notably those of Vaucanson, whose mechanical serpent could hiss and dart on Cleopatra's breast (a particularly imaginative fantasy for a machine, blending, sex, queenly power, and fear of death). From such imitation machines, La Mettrie concludes that Vaucanson's skill points to the undoubted possibility, in the workshop of the next Prometheus, of the making of a talking man (140–41). Here La Mettrie demonstrates the contemporary enchantment with these robots and the excited elevation of the machine to a paradigm for humanity and the universe itself: "I am right! The human body is a watch, a large watch constructed with such skill and ingenuity" (141). In this large watch, he imagined the heart to be "the mainspring of the machine" (141).

One problem that such mechanists had to answer was the source of motion. If the universe is nothing but mechanical matter, how did its motion originate? While older thinkers such as Aristotle postulated God to be the first mover, La Mettrie represents the emerging *elimination* of the divine from the clockwork and the relegation of the question of origins to the trashbin of unanswerable questions: "The nature of motion is as unknown to us as that of matter. . . . Grant only that organized matter is endowed with a principle of motion" (140). Choose your unanswerable questions.

He similarly brushes aside other religious questions, such as the question of immortality, to make room for humanoid machines:

> Let us not say that every machine or every animal perishes altogether or assumes another form after death, for we know absolutely nothing about the subject. . . . What more do we know of our destiny than of our origin? Let us then submit to an invincible ignorance on which our happiness depends. (147)

These questions he can ignore, but La Mettrie is truly enchanted when he fantasizes mechanical man's positive utopian potential. We will be happy as robots, he imagines, able to feel nature and aware of being part of the "enchanting spectacle of the universe," and we will *never* destroy nature or other beings (147–48). Robots will be so full of humanistic inclinations that they will even love their enemies, as Jesus taught. They will follow the Golden Rule by doing unto others what they would have done to them (La Mettrie 147–48).

La Mettrie's mechanistic thought overflows with utopian dreams. Astonishing as it seems now, in light of the twentieth-century's destructive mechanized wars and ecological crisis, he imagines a fantastic robot full of spiritual ethics of unknown origin in a machine. His heavenly mechanical humanity, with absolutely no causal, mechanical explanation, would rise far above most human ethical behavior. What in the mechanical world would make a machine act ethically? He brushes aside questions such as the origin of mechanical motion and dashes in with wild speculations about the mysteriously ethical refinement of machines. Blind to the radically nihilist implications of his philosophy, which later developed in arenas such as Robot Wars, La Mettrie naïvely trumpets wild achievements for future materialist mechanism in the name of the successful methods of science.

> Let us then conclude boldly that man is a machine, and that in the whole universe there is but a single substance differently modified. This is no hypothesis set forth by dint of a number of postulates and assumptions; it is not the work of prejudice, nor even of my reason alone. . . . Experience has thus spoken to me on behalf of reason. (148–49)

La Mettrie's simple and direct assertion of the mechanistic theory in 1748, with its denial of broader questions and its utopian dreams, is an early expression of the continuing enchantments of robot mythology's Pinocchio Project—playing with mechanical philosophical puppets.

Metallic Imagination

This android robot of the mythic realm has emerged as a classic image, the anthropomorphic icon of the technological worldview, elevated far above the level of a mere tool. The anthropomorphic robot has become a key symbol, the sacred machine par excellence for an otherwise rationalistic consciousness. It is revealing, however, that, in La Mettrie's confused mixture of materialism and other philosophies, he discards reason and envisions all-pervasive imagination as an interior projection, the absolute behind his mechanistic philosophy. Contrary to all his mechanistic visions, he says, "I think that everything is the work of imagination, and that all the faculties of the soul can be correctly reduced to pure imagination" (107). If La Mettrie is right about the basic role of imagination, he neglects to apply this insight to his own imaginative assertions about mechanism, which, according to his own rule, must be more imaginative than mechanistic.

Early twentieth-century futurists wrote plays using characters that they called "electric puppets," especially in the Italian *Teatro Grottesco* (Theater of the Grotesque), around World War I. The Italian futurist F. T. Marinetti published a play *Poupées Electriques* (Electric puppets), staged in Paris in 1909. Two metallic android puppets that looked like robots appeared in his first performance, which caused an uproar of hisses and laughter. A second performance was entitled *Electricità Sessuale* (Sexual electricity), since Marinetti dreamed of female sexual energy animating his puppets (Segel 260–63).

Enrico Carvacchioli's plays had a character called "the mechanic" who makes speaking puppets resembling real people (Segel 283). He then creates a lifelike puppet to imitate a girl thought to be dead, to comfort her grieving parents. These playwrights intended their animated puppets to be a critique of what they saw as their artificial, alienated bourgeois society.

Karel Capek wrote anti-utopian plays such as *Ex Centro* (1912), in which a character says, "Man is God in miniature, therefore he creates." In one of Capek's plays, electric puppets that look like real women are portrayed as becoming human, during romantic entanglements with men who dream of compliant, unresisting mistresses (Segel 300–301). Capek's classic play/film *R.U.R.* (Rossum's Universal Robots) in 1920 took his critique of artificial mechanisms envisioned as becoming human much further: Robots made for labor rebel, destroy humanity, and evolve into humanoids that can love and reproduce. For Capek this was a terrible apocalypse.

In 1975 the Polish playwright Tadeuse Kantor saw robots as embodiments of the dark, criminal, and rebellious side of human inventiveness and was terrified at the nihilism of the urge to express life through lifeless machines. For him mechanical mannequins called robots are their inventors' way of toying with death (Segel 325–27).

Apparently the worldview of a thoroughly dead, neutral cosmos filled with soulless machines is woefully inadequate for technoculture. Robot mythology cannot resist imagination and envisions that the machines have come alive. Samuel Butler saw in the nineteenth century that machines had turned into ersatz love objects, fetishes, and almost gods (Mumford, *Pentagon*, 196). While the surface fantasy is that mythic robots symbolize the future, this is not their unconscious message. Rather, they symbolize the present fanciful foundations of a purely mechanical world. The android robot is imagination taken metallic form. These robots are imagined with totally unnecessary human appearances (Menzel and D'Aluisio 129–31). As Baudrillard says, imagination takes priority over technique: Automatic machines fascinate so powerfully precisely because they are *not* rational. Rather, they enchant because of the imaginative force embodied in our robotic creations. Our compelling desire for automatism—for everything to work by itself—is a new anthropomorphism, embodying our unconscious desires in machines (Baudrillard 111). This grants collective permission to imagine machines wildly.

In this fantasy world of metallic vision that pretends to be a future possibility, robot mythology elevates objectness to a transcendent absolute form that is capable of autonomy. The dead world of mechanism magically seems to come alive with exciting objects acting like subjects. Baudrillard spots illusion. Automatic machines serve rhetoric and allegory, so the fantasy gadget is not at all "objective," he charges, but rather a delusory projection of consciousness (113).

But can't sensible, rational minds distinguish objective reason from imagination when encountering a robot? Of course they can, yet they are still fascinated. Cynthia Breazeal, a robot scientist at MIT who constructed an anthropomorphic head with big "eyes" and red "lips" named Kismet, says:

> "To me, the ultimate milestone is a robot that can be your friend. To me, that's the ultimate in social intelligence. . . . We anthropomorphize all kinds of things, our pets, our cars, our computers. . . . With an android, as with a good story, there must also be a willing suspension of disbelief. . . . [W]ith Kismet they tend to say, 'It smiled at me!' or 'I made it happy!'" (qtd. in Whynott 69, 72)

At some point robots and humans will coexist, Breazeal says, "'And they won't just be appliances; they'll be friends'" (qtd. in Whynott 69–72). Humans are easily enchanted by our doll-like creations. We want relationships in various forms, so we build them.

Anne Foerst, a theologian and roboticist, who stresses that empathy prevents objectivity, expresses the paradox of highly rational robot engineers interested in giving machines human qualities. She rightly stresses that in building machines analogous to humans, one makes assumptions about human nature:

> "Is the robot Kismet a she?"
> "Robots are its. But I cannot help but think of her as a she. . . . Of course part of you thinks, It's just a stupid machine. But you do react and you can't help it" (Dreifus "Androids," see Foerst 60).

Here is the circular logic of the Pinocchio Project. Human nature is not simply mechanical but imaginative—enchanted—and that same imagination wants to make mechanical puppets seem human, so we can imagine humans to be nothing but mechanical. The enchantment with android robots is easily triggered by imaginative qualities that we choose to build into them, such as Kismet's moving eyes, mouth, and ears. These materialized analogies, puppetlike partial simulations, are made to seem like human forms.

Artificial intelligence is analogous to the mind as a hammer is analogous to a fist. We can playfully allow such analogies to soften reason into a mystique that floods rational analysis like water in a living plant. We *imagine* robot actions to be behavior and feeling because we willingly make it look that way, then regress like a child playing with a doll, or relax into the suspension of disbelief of adult fascinations with an archetypal film character. Interacting with machines such as Kismet (sounds like "kiss me"), we expect the imagination to be under the control of reason and will, but it is a slippery character. As soon as we feel charmed, feel fun, feel as if we are being drawn into a human relationship, we are in emotional and imaginative, not rational, territory. It is easy to be tricked into thinking android robots or good puppets are human, but this is a theatrical, not scientific, consciousness. Robots can only astonish us when we ignore the mechanism. They are not at all autonomous, since we design them, and so their functioning must not be taken as behavior.

Perhaps remembering the mechanism behind the relationship is not really under the control of rational will power. Indeed, just as imaginative themes float into consciousness unexpectedly in daydreams, robot themes seem to guide thinking more than we realize. Why would we want to make impractical android robots and numerous mythic books and films about them if they were not imaginatively captivating? As Freud showed, the conscious will is not autonomous but is influenced significantly by unconscious dynamics.

Another highly imaginative quality of robot myths is sex. Why are robots sometimes pictured as sexy women or men? What does sexiness have to do with their mechanical function? The classic robot imitating Maria in *Metropolis* has a lovely sculpted body with feminine curves that often reappear in robot art. The musclebound weightlifter Arnold Schwarzenegger, who plays the robot in *The Terminator,* is also a sexy male image of seductive, muscular machinery. The film's director, James Cameron, said to Schwarzenegger about his role, "'You are a machine, I'm telling you Arnold, you are the symbol of power and like a machine—a fine-tuned machine. . . . You'll be a very memorable cyborg villain and a human machine'" (qtd. in "Body" 74).

Some dreamers foresee robotic sexbots, like the gigolo Joe in *A.I.* Joel Snell predicts, "Robots that provide sexual companionship are likely to become common in the future" (371). Reportedly, prototypes have already been developed in Japan. "The future 'sexbots' will have humanlike features and will be soft and pliant. . . . Marriages may be destroyed by sexbots. . . . Robotic sex may become addictive. . . . Robotic sex may become 'better' than human sex" (Snell 371). When mechanical sex becomes a robotic art form, you know that technology is slavishly serving desire. The gratuitous feature of sex built into science fiction and robots is apparently meant to engage the erotic imagination in an enchantment that lures it into feeling technical machinery and its metaphysic to be compulsively seductive. We make our machines sexy so we can feel sexy with them, while fantasizing that sex can be reduced to mechanical functions, like fairy-tale Pinocchio puppets magically transformed into "real" humans. The archetypal sexuality of robot myths expresses a yearning to control that passionate organic aspect of reproduction by making it into a machine. Is this emotionally healthy? The feminist theologian Mary Daly calls the robot mechanization of life *the* patriarchal pathology, the major sickness of masculine-dominated technologies (61).

Cybermetaphysics

Why not just let robots be machines instead of re-creating the Pinocchio story in computerized form? Why would Isaac Asimov and his audience focus on the fantasy of a machine's gaining the capacity to feel and love? Why is industrial culture obsessed with whether robots can feel and become humanoid? Why would not dumb robots doing limited work be sufficient?

The underlying purpose of this fictional play is metaphysical. It is not only to fantasize that robots can become human but to advocate the converse, a metaphysical claim central to modernity's technological project. This is the metaphysical claim that humans are nothing but machines, in the tradition of Descartes and La Mettrie. This is the experiment: If a robot can succeed in becoming human, with the help of a mythic robot engineer standing in for Dr. Frankenstein, then this seems to prove that humans can be reduced to machines. If a machine can be made to replicate a human, the enchantment goes, then humans can be explained as only machines. Of course, there are mechanical aspects of humanity, such as bone-joint movement, but that cannot be extended to explain the vast complexity of humanity. The hubris of this immense Pinocchio pretension is illogical and incredible, but in mythic form, it seems plausible. The repetition of androids in cinema, television, and computer games gives apparent personalities to fantasy robots that are actually nothing but theatrical tricks. Andrew in *Bicentennial Man* and C3PO in *Star Wars* are inhabited by actors, and, of course, Data on "Star Trek" and David in *A.I.* are human actors pretending to be robots.

The concealed metaphysical claim is not a logical argument but an enchanting drama. It neglects and denies a mountain of difficulties in analyzing human nature, from reproduction to feeling. The mechanical metaphysic has succeeded in building so many powerful machines, from steam engines to rockets, that it boldly dreams of universal application. Mechanistic metaphysics fantasizes that it can explain human nature mythically in newly computerized versions of Pinocchio. The attempt to grant robots feelings is the frontier of this metaphysical probe, but its only convincing examples are cinematic tricks.

In the film *A.I. (Artificial Intelligence)*, a dismal future includes robots that have been "programmed" with feelings. Another cybernetic version of Pinocchio motivates a robot "boy," David, in its quest for its lost mother. On the surface this Freudian Oedipal plot motivates the film—a robot wrongly "wounded" by sibling rivalry is abandoned by its mother in favor of her real son. But the robot "loves" and relentlessly seeks its mother. David believes that if it were transformed into a real boy, like Pinocchio, it would be accepted and loved by its mother.

The key fallacy in this metaphysical-theatrical experiment is the way we anthropomorphize machines. There are no such things as machine eyes, legs, arms, brains, feelings, love, gender, intentions, or any such human qualities. These are all human dreamlike analogies projected onto clockwork puppets in an unconscious fairy tale. This is the mythic vision of the success of an experimental puppet show, persuading the audience that the puppets are real, not just an enthusiastic, enchanted human imaginative entry into a good story on stage or screen. This psychological phenomenon, called the suspension of disbelief, takes place in that wide, fuzzy border between reality and imagination every time we get drawn into a good drama.

The film *A.I.* is unusually and ironically explicit about being a fairy tale. The abandoned robot "boy" desperately seeks its transformation by finding the Blue Fairy of the Pinocchio folktale, who, it believes, can make him a real boy and thus loved by its mother. How many science-fiction stories openly acknowledge that they are fairy tales? This candid fairy-tale quality, not the Oedipal theme or the portrayal of robots with feelings, is the most important theoretical element of the film because, unlike most robot tales, it acknowledges the imaginative quality of its portrayal of humans as robots. This is the larger Freudian downfall of the experiment—seeing the unconscious fairy-tale dimension of the whole story.

In an apocalyptic future world, the film proposes, global warming has melted ice caps, leading to massive flooding of major coastal cities. New York City is pictured as destroyed by the flood that leaves visible only the tops of useless skyscrapers. The reduction in available land forces survivors to control population. So a clever robot engineer designs a child robot that can be programmed to love human mothers deprived of real children. The godlike pose of the robot designer, Allen Hobby of Cybertronics Corporation, is captured in his line, "Didn't God create Adam to love him?" Placing this thought experiment in the mythic realm of the future echoes a classic theme of science fiction: the future Earth has been ravaged by technological excess.

Of course, criticism of technological excess is an important theme in science fiction, which ambivalently portrays speculative, optimistic successes of technology and pessimistic, dangerous failures. Industrial society's solution, however, is not to restrain current technological excess but to press forward with yet another questionable high-tech project, because technology is deemed absolute. The apocalyptic shadow of technology is evident, but its victims cannot escape the repetition of a basic fallacious enchantment of industrial culture—*more new technology will solve technology's problems.* The unchallenged absoluteness of this a priori enchantment is highlighted by the roboticist's comparison of himself to God making Adam.

The Pinocchio Project vision that robots could be programmed with human feelings that would make them almost indistinguishable from people is offered with no explanation. Faithful to the mechanical metaphysic, the film imaginatively assumes that it will be possible and forges ahead to explore the potential consequences in the love department. A child robot seems disarmingly cute and charming for frustrated parents. A gigolo robot seems to be a seductive lover for a lonely single woman. The major surface question is, What ethical obligations would humans have toward such machines? If a machine is capable of simulating love, is its owner obligated to treat it as a loving human? In *A.I.*, not surprisingly, humans do not treat robots any better than they treat other humans. Cynical children taunt them, conflicted parents neglect and abuse them, brutal circus crowds torture and destroy them. As one reviewer says, "Fear is the underside of enchantment, and the spell of wonder *A.I.* casts is tinged with dread" (Scott).

At the end of the film, the robots get their revenge. Humanity has degenerated into an ugly, brutal, nihilistic, commercial society whose technological miracles do not prevent them from slow self-destruction. The gigolo robot Joe explains to David, "They made us too many, too fast, and when they are gone there will be nothing left but us. And that is why they hate us." While cynical about human nature, this plotline grants robots a magical survival potential and ability to reproduce. Highly rational humans have failed to master their own feelings and ethical obligations to each other and to ecology, and the result is their demise. Yet robots magically rise above these frail human limits and emerge as seemingly noble machines, not only capable of brilliant technological advances but also of compassion and even some type of immortality. Machines supersede the humanity that made them, thus denigrating the emotional and spiritual aspects of human nature, which the mechanistic metaphysic has already dismissed to the "subjective" dustbin but granted to robots. Reason is transferred from humans to machinery made by humans. Rationality becomes an autonomous puppet master running around making real human feelings look inferior. This dissociates reason from feeling, which worsens the pathological situation.

In a fairy-tale leap, the film has the "boy" robot descend like Jonah in a whale-shaped flying/diving machine under the sea, where it finds a carnival Blue Fairy. It keeps begging to be transformed, until a new ice age freezes over the former New York. This mythic scheme, echoing the classic hero's journey to the underworld like Jonah and the whale to meet the divine, leaps forward two thousand years, when a new generation of robots, curiously looking like our current fantasy of wise, gentle space aliens, resurrects it and helps it partially fulfill its wish. This implies that robots are possessed of a high-tech version of immortality that exceeds human frailty and foolishness. Engineered rationality, that is, can survive far longer than mere human flesh and, endowed with superhuman wisdom and compassion, can excel where mere humanity failed. Why are humans with real feelings portrayed as so crude and cruel, while future artificial intelligence is endowed with such fairy-tale brilliance, sympathy, and ethical integrity? This fantasy pathologically gives up on any humanistic improvements for humans yet grants to alien-looking rational machines two thousand years from now all the emotional and spiritual strengths of angels.

This mythic yarn has nothing to do with objective reality, of course, but in its quest for meaning it ironically discloses the fanatic obsession with the very subjective feelings that objective rationality has banned from its discourse. Of all the powers that we could imagine future robots to possess, such as exploring remote planets, it is striking that so many robot enthusiasts are enchanted with the notion of programming such refined human feeling into robots—the very feeling that is repressed, denied, and dissociated from technorationality. Today's real robots fall far short of adequate humanlike behavior, but that does not stop the fairy tale from persisting. Is this fairy tale a compensation for our imagined nihilistic worldview? As one reviewer of the film pointed out,

> Kubrick obsessed over *A.I.: Artificial Intelligence* for nearly two decades, at one point experimenting with an actual robot covered by plastic skin that he hoped could star in the movie. "The robot was lifeless—a total failure," [co-producer Jan] Harlan recalled with a laugh. "It really looked grotesque, and we gave up on it." (Breznican 7)

This master storyteller pressed on, even though the basic premise of his story—that a robot could be a convincing boy—seemed so impossible.

This kind of fairy tale, which even delights in its fairy-tale-ness, reveals that the whole Pinocchio Project about robots having feelings and exceeding human nature is very important. But the theme of emotions dissociated from reason is wrong. The Pinocchio Project shows that emotions such as fear, love, and anger are constantly embedded in our rationality—not separated, not illusory, as the fairy tales portray in industrial culture, but right up front for all to see. The fairy tale of robots capable of feeling, sexual behavior, dignity beyond humans, resurrection, and possessing immortal souls is an archetypal, spiritual enchantment right at the heart of the Pinocchio Project and high-tech culture's rationalism. Archetypal spirituality is driving the machine, right next to the Nietzschean will to power, the Freudian lust for pleasure, and the fear of a meaningless, nihilistic existence.

Desire is not a denied subjective distraction from objective reason; nor can it be reined in through the application of more reason. Passion is thoroughly blended with logic, even driving or overriding rationalistic thought with its dreams, fairy tales, and myths. The master desire driving robot mythology is the dream of absolutizing the human subject's rational metaphysic to the extreme that its robotic products exceed humanity itself. This fantasy extension of rational engineering breathes enchantment and dread.

Magical Icons

Any decent robot representing a metaphysic should be able to perform transcendent feats. As Arthur Clarke commented, "'Any sufficiently advanced technology is indistinguishable from magic'" (qtd. in Menzel and D'Aluisio 19). Peter Menzel adds, "Today the technological magic is more than accepted, it is expected. We believe that our dreams are not just dreams but are sneak previews" (Menzel and D'Aluisio 19). Our ability to engineer these magical icons seems to elevate their creators to godlike status. Robot designers are possessed by a "fascination with playing God" (Menzel and D'Aluisio 18). In the excitement over new inventions such as telegraphs, trains, and electricity, nineteenth-century questions about how mechanisms could explain mysteries such as consciousness, feeling, and morality were brushed aside as romantic sentiments. By the twentieth century, computerized robots that enabled more simulations of consciousness became powerful icons of transcendent dreams. As Jean Brun puts it,

> In this [Cartesian] epic, much more metaphysical than epistemological or technological, the construction of android robots represents a decisive stage in . . . [the quest] to achieve self-transcendence. . . . the fabrication of the robot . . . is guided by a demiurgic desire. (Brun, *Masques*, 77–78)

The engineering of androids is driven by this demiurgic dream of divine creativity far more than technical utility.

Technically minded thinkers may believe that technology is the enemy of magic and religion, but robot literature suggests that the opposite is true. While the successes of the machine age seems to have been weakening the credibility of the traditional gods of past eras, high-tech culture has been developing its own theology. The ontological thirst of the industrial mind, its search for the ultimate nature of existence, has not been suppressed by the mechanistic worldview in its attempts to eliminate traditional gods. Instead, skeptical science has given birth to magical icons of godlike powers.

Obviously, modern robot mythology runs deeper than fictional speculation and forecasts of the future. Not only have robots come alive and gained souls in the technological imagination, they have become symbolic of a particular kind of theology. The robot has become a central symbol for the transfer of the sacred to the technical, as described by Jacques Ellul:

> Nothing belongs any longer to the realm of the gods of the supernatural. The individual who lives in the technical milieu knows very well that there is nothing spiritual anywhere. But man cannot live without the sacred. He therefore transfers his sense of the sacred to the very thing which has destroyed its former object: to technique itself. (143)

The Pinocchio Project has become a quest not only for earthly but for *absolute, universal* mastery by technological consciousness. As Gary K. Wolfe notes, "robots have taken on the aspect of a mechanical god" (168). The mechanistic metaphysic that loudly threw spirituality out the front door quietly lets it in the back door. Thus there is a great absurdity in the machine worldview's anxious attempts to create robotic life-forms. The effort to invent lifelike automatons is an effort to create not only useful machines but life itself. Success at Dr. Frankenstein's tormented goal would seemingly demonstrate that whoever can create life is a god.

Godlike Robots

The themes of robots and robot makers being seen as gods is an expression of the dramatized quest for ultimate reality. Heinrich von Kleist, whose thought stimulates many actors and puppeteers, wrote about grace as a gift of the divine:

> [G]race itself returns when knowledge has as it were gone through an infinity. Grace appears most purely in that human form which either has no consciousness or an infinite consciousness. That is, in the puppet or in the god. (Kleist)

When holiness is unacknowledged but unconsciously present it has eerie characteristics that promote the suspension of disbelief and lure the unsuspecting soul into a mythic world through imagery that makes an end run around conscious reason and speaks straight to the unconscious soul. Thus drama becomes spiritual enchantment, through a child's beloved doll, a spooky Egyptian mummy, a seductive statue of Aphrodite, or a powerful robot. The embodiment of the spiritual, or "grace," in ancient frescoes, statues, icons, puppets, and now robots has an uncanny quality to it, as Victoria Nelson observes in *The Secret Life of Puppets* (31). When stone, wood, or steel are sculpted into anthropomorphic images, they draw the soul down, below the subject/object worldview, deeper into the primordial flowing stream of existence itself, Being, or your face before you were born. This is a fascinating and uncanny experience, known in its raw, preverbal form to mystics of many stripes.

For most of us Being is known in images that allow fragments to sneak through icons of gods or goddesses, heroes, burning bushes, virgin births, resurrections, and now in technologies that reach for the stars and evoke awesome wonder at seemingly transcendent feats—robots hurtling through vast distances through space and sending back photos of incredibly vast and dazzling starlit clouds of wonder. Mythological android puppet robots perform miracles—going beyond time-space, reproducing themselves, and overcoming death. The simulacrum is often imagined to be equipped with superior ethical and spiritual qualities than humans. The robot as puppet seems to steal individual freedom from humans by gaining divine status, as in the alien-looking robots at the end of *A.I.* We mere humans project our divine spark onto grandiloquent robots, as in *Metropolis* or Ray Kurzweil's thought. One British thespian urges actors to become "*Über-marionettes*," transforming themselves into a "body in a trance," thus becoming the archetypal double, as in robots (Nelson 256).

As the clone or homunculus, puppet robots carry our hidden traits. We project our unconscious archetypal energies onto the seemingly autonomous puppets that we call android robots. They enact all sorts of enchanted dramas—villains, murderers (*The Terminator*), illusion-creating magicians (*The Matrix*), lovers (*A.I.*), children (*A.I.*), tyrannical masters (*Metropolis*), slaves (*The Stepford Wives*), heroes (*Star Wars*), or victims (*A.I.*).

As puppets, android robots are full of paradox—unfeeling offstage, but kind, seductive, charming, loyal, or obedient onstage. They are ugly machines inside but can be beautiful, even sexy externally (*Bicentennial Man*). Robot puppets are powered not by magical battery packs but by our unconscious imagination. We call machine parts "eyes," "ears," "brains," "arms," "legs," and imagine them to possess human thoughts and feelings, projecting human qualities onto machines that we have built and programmed just like we do with puppets. Robots are more autonomous than wooden or cloth puppets, since they can operate without hidden hands or strings, unlike puppets, and this is more enchanting to us, but our interaction with them is very similar.

Most important, we project onto our puppet robots and their makers the divine wonder and power that our proposed mechanical worldview denies to us. The fantastic, transcendent powers of the highly imaginative android robots in our industrial mythology turn lifeless matter into apparently living Pinocchios that can conquer evil, overcome time, space, and death, and even sacrifice themselves for our love (*Bicentennial Man*). This provides a stimulating drama for the ongoing development of mechanistic and electronic metaphysics and seems to elevate robot makers to divine status themselves, since they can create simulacra of living beings, so they boldly indulge in wild speculations about future robot potentials. However, this playing with divinity is an uncanny, thus fascinating, compensation for the real perceptions of divine light in real humans that has been cast aside in the mad, enchanted rush of industrial society, and a dangerous quest for absolute power.

The Ontologically Creative Robot

Like Michelangelo's God about to touch Adam's finger and infuse him with life, creators of robots assume a godlike generative power. Mary Shelley's Dr. Frankenstein is the classic image of the modern Promethean divine creator. He not only has the fantasy skill to assemble the bodily parts of the humanoid creature, but he also has the transcendent ability to animate the corpse. We can see in the older legend of the Golem a myth of the power of the divine creative word, as Jean Brun points out. In that sixteenth-century legend, the Golem is a clay android that comes alive when Elijah of Chelm writes the secret name of God on its forehead. This is a mythic reenactment of the divine creation of life, which is now claimed by technological dreamers (Brun, *Masques,* 77).

Similarly, in Fritz Lang's *Metropolis,* an industrial tyrant who has enslaved his futuristic city's robotlike workers employs a Promethean inventor to create a destructive robot in his magical-mechanical workshop. It is disguised as the lovely Maria. Before her capture, the real Maria had been the workers' spiritual leader and was planning a peaceful revolt. The destructive robot, masked as the virginal/motherly Maria, brings massive deception and destruction. Many references are made to divinities in the film, including the ancient Nordic goddess Hel as the goddess of the underworld and death and the devouring Ammonite god Moloch.

High-tech culture can manipulate life's seeds, but actual creation out of dead matter exceeds its grasp. Nevertheless, the "artificial life" movement dreams of this absolute power with shallow definitions of "life." And already another leap has been made, from the portrayal of humans creating living robots to the fantasy of robots that create their own kind. In Asimov's *Robots of Dawn,* a scheme for robots that can reproduce robot babies is seriously proposed (287). Mythic robots join the ranks of gods as saviors, avatars, and saints, those semihuman, semidivine figures in every religion. This is a magical machine that is not a machine, a powerful human that is not a human, a mechanical god that is not a god. The theology of industrialism is not comforting. This is because the magically creative and out-of-control destructive anxiety behind Pinocchio myths cannot assure a central theme of world religions, which is the ultimate strength of divine cosmic compassion and justice to overcome suffering, injustice, and death.

The Savior That Destroys

The divine machine is both benevolent and destructive. This is evident in the *Star Wars* saga, in which Darth Vader converts from the light side of the Manichean Force to the dark side. Like two schools of theology, there are two types of robot speculation, with each position admitting aspects of its opposite: the savior robot has a cold and dangerous shadow double, and the destructive robot has a fascinating appeal.

The optimistic view of the robot as savior was well represented by Isaac Asimov, who claimed that robots could be controlled by a carefully programmed moral code called the Three Laws of Robotics. Like a postbiblical Three Commandments, these will always—he hopes—keep robots under control and prevent them from injuring humans. They are:

1. A robot must never injure a human.
2. A robot must obey orders from a human, unless they conflict with Law 1.
3. A robot must protect is own existence, unless this conflicts with Law 1 or 2. (Asimov, *I, Robot,* 38)

His robots thus become idols—golden calves of the mechanistic worldview, symbols of the mystique of the ultimate benevolence of technology.

This optimism extends to assumptions for research, such as those expressed in Asimov's *Robots of Dawn:* "If you know how a robot works, you've got a hint as to how a human brain works" (252). This model blithely reduces the human soul to a computer and then reduces morality to three programmed Laws of Robotics. In his optimism, Asimov imagines the real danger not to be robots but the failure of humans to be rational. In *I, Robot,* for example, a mother harmfully denies her daughter a beloved robot babysitter. Asimov sets this up to seem irrational on the wild assumption that robot babysitters will be thoroughly, rationally android. Asimov must have actually found his imaginary robots' rationalism boring, however, because he was fascinated with human passions and religion. He promoted the argument that the computer model of the mind will explain even passions and mysticism. The techno-optimists worship robots as mechanistic, yet in some unexplained way also as passionate models of human mind and soul—the Pinocchio Project.

One captivating expression of belief in the benevolence of robots is the apparently irresistible tendency to make them cute. The Tin Man in *The Wizard of Oz* is one of the most attractive early mythic robots. Its song and dance could easily be named "The Hymn of the Charming Robot." Its appealing lament—"If I only had a heart"—is so delightful that it makes the cultural emptiness that it symbolizes almost tolerable:

> The tinsmith forgot to give me a heart . . . All hollow . . . If I only had a heart . . . Just to register emotion—jealously, devotion—and really feel the part . . . If I only had a heart.

Despite its human form and face, its captivating song and dance, the heart is lacking in the hollowness of the machine. Modern descendants of the singing, dancing, heartless robot are seen in C3PO and R2D2, which have stimulated a succession of robotic toys, talking teddy bears, and aliens. These robots not only sell enchantment and affection but offer reassurance that such cute machines are ultimately benevolent.

The skeptical view of robots rejects such starry-eyed optimism. In *Robocop*, the corporate greed of a degenerate industrial society stimulates the misuse of robots. The giant gunslinging kangaroo-dinosaur robot that accidentally murders a businessman is glibly tolerated by a corrupt executive. This myth teaches that the potentially benevolent mechanical creation only turns destructive when misused by humans. The sympathetic monster-robot image teaches the instrumentalist view that if the kindly machine is transformed into a demonic monster, humans can only blame themselves. This view imagines robots as mere tools, subject to rational control, but it naïvely neglects their symbolism.

Far from being useful tools, anthropoid robots are the unconscious externalizations of human fantasies, philosophies, and theologies. Android robots seem to be highly rational machines, but in fact they emerge from unconscious enchantments. They unconsciously guide dreamy schemes for making more machines, benevolent or destructive. And once they are released into operation, they may have a nightmarish autonomy.

For example, in the mythic film *Westworld*, the stone-faced cowboy robot, played by Yul Brynner, somehow tires of its programmed victimization, established merely to satisfy every egotistic and lustful human whim. When the entire system mysteriously collapses and the programmers are destroyed, it rebels. The robot cowboy then pursues one human who had shot it once too often. The robot survives defensive assaults of acid and flame. Resurrected from these deadly attacks like a divine hero in classical mythology, it almost succeeds in its demonic revenge before it collapses, a burned-out hulk of metal and wiring. In this story, despite the best of human control and intentions, the apparently benevolent machine absorbs human passions and becomes *inherently* destructive. The passions and powers that are built into technology can too easily overwhelm the instrumentalist control in which Asimov had faith. This pessimistic theme expresses little faith in the machine worldview and its consequences; rather, it shows that the rationalist premises of the optmistic technological mental framework are tragically defective. The apparently benign android robot may be a traitor in our midst.

The Brain as a Machine

Just below the rational level of argument about computers and their "artificial intelligence," the unconscious influence of imagination and the quest for absolutes is at work. One effect of this is argument by assertions of a priori assumptions and circular definitions rather than by careful reasoning. For example, Marvin Minsky often simply assumes that the organic human mind is at base a machine. Organized in a hierarchy or "society" of various mindless small parts, he sees at the bottom only mechanisms, the only other alternative being "magic." Like many technologists, he places universalizing faith in the premise that physics can explain the world: "The science of physics can now explain virtually everything we see, *at least in principle,* in terms of how a very few kinds of particles and force-fields interact. Over the past few centuries reductionism has been remarkably successful" (Minsky, *Society,* 26). This classic materialistic reductionism rejects the more humanistic, intuitive gestalt or the holistic theories that Minsky hastily dismisses as "pseudo-explanation" (27). But the collective enchantment with materialism of the mind may be too hasty in accusing other views of being "pseudo" and "magic." The common psychological defense called projection may be at work here, in which we accuse our opponents of the very problems ("pseudo-explanations") from which we unconsciously suffer. But playing with magical pseudo-explanations is not only a personal problem for Minsky; it is the collective Pinocchio Project in which he participates.

Minsky undertakes to show that "you can build a mind from many little parts, each mindless by itself" (*Society,* 17). He builds his argument with numerous engineering concepts, based on the a priori assumption that humans are machines. Of course, if you assume that, then no matter how clever or faulty the intervening logic, you can easily conclude what you assume. Responding to the criticism that he reduces the mind to a machine, he replies with curiously circular arguments. First, he uses a slippery fallacy: "[M]ost people say, '*I certainly don't feel like a machine!*' But if you're not a machine, what makes you an authority on what it feels like to be a machine?" (30). This is like the fallacy, "If you have never seen a ghost, how do you know there are no ghosts?" This presumes that there are ghosts, just as Minsky presumes that we are machines.

Then Minsky tries to evade the issue by saying that the "machine" word is outworn: "the word 'machine' is getting to be out of date. . . . The term 'machine' no longer takes us far enough" (30). There is an interesting thought, given the electronic nature of computers in contrast to, say, trains. However, computers are still machines, not organisms. Then he calls on the old progress enchantment, which is pretty weak, given the number of historical technological failures and disasters: "We cannot grasp the range of what machines will do in the future from seeing what's on view right now" (30). But lack of information is not information. Finally, as he attempts to build a foundation for his argument, he slips the machine image back in, after trying to dismiss it: "Let's put these arguments aside and try instead to understand what the vast, unknown mechanism of the brain may do. Then we'll find more self-respect in knowing what wonderful machines we are" (30).

What? The machine concept is outdated, but what wonderful machines we are? This circular argument blatantly undercuts itself by failing to demonstrate that we are machines, while insisting on *assuming* so. Such fallacious rhetoric may be expected from an amateur robotics enthusiast but not from a scholar supposedly trained in logic. Minsky's theory is that at base the mind is a society of mindless parts. This is another leap of fallacious thinking bordering on magic, assuming that a large collection of mindless things can somehow acquire mental powers. This is analogous to assuming that a vast field of junk cars can, by some unknown technique, assemble themselves into a jumbo jet and take off—or that the more chaos your room contains, the more likely it is that it will clean itself up into a tidy, meaningful order. The enchantment here is a strong one, able to leap logical barriers and convince Minsky and other roboticists that the mind is a machine, a computer that is at base mindless:

> It was only after trying to understand what computers—that is, complicated mechanisms—*could* do, that I began to have some glimpses of how a *mind* itself might work. . . . How could we ever have expected, in the first place,

to understand how minds work until after expertise with theories about very complicated machines? . . . In years to come, we'll learn new ways to make machines and minds both act more sensibly . . . learning for ourselves to think of 'thinking,' 'feeling,' and 'understanding' not as single, magic faculties, but as complex yet comprehensible webs of ways to represent and use ideas. (Minsky, "Why," 165)

Minsky asserts that there is no problem with such circular definitions, since we will just write programs for circular definitions. He sees no mystery to the world, and indeed says, with incredible materialistic hubris, "Look, the world is a rather dumb place. There's nothing special about it. It's accidental. The world was *terrible* before people came along and changed it. . . . Eventually the robots will make everything" ("R.U.R." 369).

This expresses one of Minsky's dominant a priori assumptions—the idea that the entire world is essentially dumb and mechanical and that human creations are the only way to improve things. Artificial intelligence and robots are the best path to that improvement. Since the human mind is basically a machine, its contents can be replicated in robots. This mechanistic definition cannot be literal, of course, since the world is organic and mechanism is an abstraction of natural structures and forces embodied in artificial machines. This metaphor of the world as a machine is pervasive, though, and Minsky is so enchanted with it that he is bolder than many robot engineers in asserting its implications, illogical as they may be. The illogical nature of his preliminary argument weakens his case for a logically structured "society of mind." Circular rhetoric and large leaps into imaginative premises such as the mind/machine metaphor simply increase his reliance on enchanted fascinations with mythic images. This technological framework of robotics rests on highly fanciful ideas of humanity as nothing but machines. Far from hard science, these illogical claims have taken on the mantle of absolute, unquestionable assumptions, otherwise known as dogmas.

Another engineer who sees the brain as nothing but a machine is Ray Kurzweil, whose dominant metaphor is that old standard, progress. He generates a graph showing the accelerating progress of computers and boldly proclaims it to be a law:

The Law of Accelerating Returns predicts that both the species and the computational technology will progress at an exponential rate, but the exponent of this growth is vastly higher for the technology than it is for the species. Thus the computational technology inevitably and rapidly overtakes the species that invented it. At the end of the twenty-first century . . . computers . . . will be vastly more powerful (and I believe more intelligent) than the original humans who initiated their creation. (*Age* 255)

Blithely neglecting the destructive shadow of technological progress in modern warfare and pollution, Kurzweil is so enchanted with the progress metaphor that he dares to make such sweeping predictions.

Kurzweil's 1999 book *The Age of Spiritual Machines* takes on this conceptual challenge of transforming machines into spiritual beings. Since he predicts that robotic machines will be trillions of times more complex than humans today, he believes that they will be capable of what we call spirituality. "Just being—experiencing, being conscious—is spiritual, and reflects the essence of spirituality. . . . [Machines] will believe that they are conscious. They will believe that they have spiritual experiences. They will be convinced that these experiences are meaningful" (153). He bases this wild extrapolation on his belief that neurologists have found the "God Spot" in the frontal lobe, the basis for spiritual experiences, developed through evolution due to the social utility of religious belief. Kurzweil thinks that within a century future robots will meditate and pray like humans. This philosophically and theologically naïve view assumes that spirituality is nothing but an aspect of an electronic brain that can be replicated by machines. This old-fashioned materialistic reductivism is criticized by philosophers and scientists alike.

The philosopher John Searle argues that Kurzweil in no way proves that human consciousness could be *duplicated,* only that it could be *simulated* by a machine. You could not duplicate real digestion by programming it on a computer either, no matter how large and fast the machine. Computers are good at simulating many things, as cinematic tricks show, but the leap to calling that "duplication" stumbles on several errors. One is the metaphorical use of the term "intelligence." Is it valid to say that a hand calculator is more "intelligent" than me because it can calculate faster and more accurately? Does such a machine "know" what it is doing? To call computer calculation and symbol manipulation "intelligence" is to neglect the wider range of human intelligence. Humans actually do fairly little computation but far more functions on a wider scale, such as making judgments without all the desired information. If you are driving down a street and see a red hexagonal sign that says "TOP," you would know from the context that means "stop," even though some information is missing. Being a good driver, you would not mistake it for a sign pointing to the top of something. Intelligence is far greater than computer calculation, no matter how large or fast. Searle says, "Increased computer power by itself moves us not one bit closer to creating a conscious machine. It is just irrelevant." There is a huge gap between the spectacular claims Kurzweil makes, Searle charges, and the weakness of the arguments (Searle 72, 76).

Furthermore, the genecticist Michael Denton argues that "there is no convincing evidence that living organisms are strictly analogous to artificial mechanical objects" (Denton 79). Living organisms can reproduce themselves, grow and change, and assemble themselves into life forms. Wonderful as fast, large computers will doubtless be, the aggregate of mechanisms will never approach the structure of organisms, any more than a huge aggregate of disorder, such as a sloppy bedroom, could autonomously form itself into an ordered system.

Machine and organism are simply two distinct categories of being. As such, organisms do not work using the identical removable parts that can all be replaced, as in a mechanism. Organisms are not completely modular. They operate on the principles of reciprocal formative influence. From proteins to organs, if you remove a part, the remaining portion will react seriously, even with death. Unlike mechanical structures built from the bottom up from basic nonliving parts, for organisms, "the form and function of each part is determined by the whole." The parts of an organism cooperate with global influences on the entire organism. The parts of a cell are existentially dependent of the whole, Remove some parts, and eventually all of a cell will die. It is because of organic "reciprocal self-formative interactions" that reductivism will not ever work as a method for building organisms like the human brain. Machines are no more than the sum of their parts in isolation, but organisms are fundamentally different. They are essentially nonmodular wholes, and holistic order is intrinsic to their functioning, more in a top-down form. Organisms possess properties that machines do not possess, even to a small, expandable degree. Organisms and machines belong to separate ontological categories.

As a result, Denton the biologist concludes: "all nature becomes re-enchanted." Genetic determinism and evolution, for example, cannot be explained on the materialistic model of life originating in small units of dead matter with no prior overall organizing system.

William Demski, a scholar of the conceptual foundations of science, argues that the materialistic foundations of the artificial intelligence movement dream of an impossible and impoverished spirituality. "If predictability is materialism's main virtue, then hollowness is its main fault. Humans have aspirations. We long for freedom, mortality, and the beatific vision" (99). Wide-eyed techno-culture looks for salvation in technologies, such as behavior modification or genetic engineering, but this quest for spiritual salvation from machines is misplaced.

Spirituality includes morality, which cannot be attributed to a machine. Despite frequent anthropomorphizing imagination, our cars, toxins, and weapons are not out to get us. They have no free agency, not even behavior. Cognitive scientists say that intelligent agency "supervenes" on natural processes. But this simply means a hierarchical relationship between lower- and higher-level processes. It makes no pretence of explaining *how* lower elements, say enzymes, causally direct the higher functions, such as consciousness. Neuroscience research is too modest, Demski says, to support materialism's "vaunting ambitions."

Materialists may scorn the psychology of beliefs, desires, and emotions without a materialistic basis as "folk psychology," but they may be seen as doing "folk materialism." Neither rhetorical term is helpful. "The world is a much richer place than materialism allows, and there is no reason to saddle ourselves with its ontology" (Demski 113). Matter is a narrow abstraction, like a car's dashboard conveniently abstracts the functions of a vehicle but in no way completely indicates the vehicle's total functioning. Materialism is too reductive to explain the entire range of intelligent human agency, much less all of reality.

How do materialists assert that the mind is nothing but the secondary by-product of the brain alone, when cases such as Louis Pasteur's are so baffling? Pasteur suffered a cerebral accident but continued his creative scientific career. When he died, survivors discovered that half his brain had atrophied (Demski 107). The problem is not so much scientific as philosophical and religious, turning on the question of the role of larger intelligent design, far broader than simple material components.

Kurzweil's dream of "scanning" a brain into a computer is a simple-minded form of the "Turing Fallacy," and its fantasy of duplicating brains in machines, argues the zoologist Thomas Ray:

> The materials of which computers are constructed have fundamentally different physical, chemical, and electrical properties than the materials from which the brain is constructed. It is impossible to create a "copy" of an organic brain out of the materials of computation. (123)

Furthermore, why would these imagined future computers, either by agreement of all their programmers or in their supposed autonomy, be so ethically spiritual and not crude? The film *Stepford Wives* portrays a nightmare where a group of computer programmers create the ideal patriarchal slave robots to replace their wives. These robots look seductive and cheerfully do housework. They cannot understand serious discussions and talk in shallow ways. In the bedroom, everything they say is to please their husbands. Once again, utopian Stepford-ites offer positive visions that neglect negative possibilities.

Real spirituality is a far greater and different phenomenon than Kurzweil's inflated electronic dreams can even imagine. True spirituality would show little interest in his mechanistic reductivism, and his notions would stimulate many amused smiles at the nearest Zen monastery. His blithe predictions, based on his happily successful technical record of inventing clever devices such as optical character readers, show no basis for understanding the ontological fallacy that he is committing by jumping from machine logic to spirituality, or the real experiences of spirituality. I would like to hear how a machine would be able to feel anger at another country for taking over its land, minimize that anger with spiritual practices, the way the Dalai Lama has, and continue teaching the compassion and peace of his spiritual practice.

Hans Moravec is another roboticist who makes bold predictions. He imagines that, as a natural extension of technological progress, robots will exceed human intelligence by the year 2050 and will eventually dominate humans. Although current computer power driving robots is about like insect mentality, and approaching lizard capacity, he predicts that it will rapidly grow to replace humans. Moravec's unique root metaphor is the image of robots as children outgrowing and replacing their parents. He simply transfers this complex organic system to machinery:

> [B]y performing better and cheaper, the robots will displace humans from essential roles. Rather quickly they could displace us from existence. I'm not as alarmed as many by the latter possibility, since I consider these future machines our progeny, "mind children" built in our image and likeness, ourselves in more potent form. Like biological children of previous generations, they will embody humanity's best chance for a long-term future. It behooves us to give them every advantage and to bow out when we can no longer contribute. (13)

Since Moravec naturally accepts the common cybernetic belief that mind is a material, mechanical organization of information rather than an ontologically unique organism, he easily imagines that robots will design minds better than human ones: "Eventually thoughtful scientist-robots may develop powerful models. . . . [B]y then their own minds, greatly expanded and improved by trial and error, will be far more elaborately organized than ours" (115). Moravec does not hesitate to dismiss mere humanity and expand his vision into outer space:

> Meek humans would inherit the earth, while rapidly evolving machines expanded into the rest of the universe. Given fully intelligent robots, culture becomes completely independent of biology. Intelligent machines, which will grow from us, learn our skill, and initially share our goals and values, will be the children of our minds. (125–26)

The archetypal enchantment with domination sees here its flip side, the expectation of submission to our own machines. This gives up on the instrumental theory's optimism about controlling technology. It is determinist and has an unconscious passive-aggressive undertone to it. We are aggressive enough to make these powerful machines, the fantasy goes, but we will finally be helpless to control them because we will make them so powerful. Here is the nihilistic strain of ultimate meaninglessness in techno-enthusiasm.

Moravec also envisions a space frontier occupied by "ex-human" cyborgs; the cost would be modest, the problems would be minimal, and they could find, exploit, and ship valuable minerals back to earth. These ex-human robot slaves (who would choose them?) will become immortal superintelligences (144–47). The old American Dream of a rich frontier with resources to exploit reemerges once we realize that robots can endure difficult space travel far better than humans. The "genius" fantasy also pops up with the "superintelligences" superior to humans, and the ancient belief in immortality blithely ignores tremendous difficulties to give family and cosmic meaning to the promises of cybernetics and robotics. This highly irrational neglect of problems suggests the need to give an absolute, almost religious significance to robot engineering. As René Dubos writes:

> [H]ow difficult it is to equate life with *known* mechanisms. In fact, to equate life with matter and its laws requires something beyond scientific imagination: it demands the *a priori* faith that living things *are* nothing but physicochemical machines. (107)

Casting away objectivity, these materialistic roboticists dream in the fanciful, dangerous fog of wishful rhetoric that envelops android robotic dreams—the Pinocchio Project victorious.

Human Intelligence in Context

Another, more limited view of the potential for robots is less dreamy. "Common sense" and understanding relevance and context, some slippery qualities of human thought, are apparently not very mechanical. Anne Foerst said in 2000, "'The most brilliant robot that the most brilliant engineers have worked on for years and years is still dumber than an insect'" (qtd. in Dreifus, "Androids"). Dave Lavery, the manager of a robotics program for NASA, says that the most advanced robotics designs still cannot get a robot to do what a one-year-old child can do easily—balance and differentiate between a shadow and a hole in the floor.

Neuroscientists are not even sure how humans generate common sense, so how can we expect to program a robot to interact that way with a changing world? This is called the "frame of reference" problem. Robots can do remarkably precise tasks in a controlled situation, such as recognizing a machine assembly error of a fraction of a millimeter. But the most advanced computers cannot approach the human ability to glimpse a rapidly changing scene, such as driving along and responding quickly, immediately ignoring irrelevant factors, to stop for a child scurrying across the road (Suplee 82–84).

Some biologists are also wary of the computer and machine models of the mind. Robert Pollack, a biologist at Columbia University, says that the metaphor of "hardware" for the brain severely underestimates the brain's plasticity. There is no permanent brain circuitry, he says, for it keeps changing and shifting locations within the brain. It is unlikely that consciousness will ever be replicated by hardware, no matter how complicated or fast (Pollack 45–46). Sherry Turkle found among computer engineers a strong tendency to project human qualities onto computers and then forget to examine their metaphors:

> Even the most technical discussions about computers use terms borrowed from human mental functioning. In the language of their creators, programs have intentions, try their best, are more or less intelligent or stupid, communicate with one another, and become confused. This psychological vocabulary should not be surprising. . . . You inevitably find yourself interacting with a computer as you would with a mind, even if a limited one. . . . "The hard part," said one [student] to another, "is reprogramming yourself to live alone." The language of computers has moved out [into the common language] so effectively that we forget its origins. . . . [It is a] discussion of minds as machines and of machines having minds. (Turkle, *Second Self,* 15–17)

Some roboticists are restrained about the potentials of robots: "To give machines the ability to reproduce, however, strikes most robotics researchers as an almost impossible task, even more difficult than building an intelligent robot. . . . To build a copy of itself, a robot would have to forage for raw materials, shape them into motors, sensors, computer chips and other parts and then put the pieces together" (Chang). Science-fiction fantasies of self-replicating robots ignore a vast number of unsolved robotics difficulties. The notion of common sense has emerged as a serious barrier for artificial intelligence. The leading artificial intelligence computer scientist David L. Waltz argues in *Scientific American* that a barrier to modeling common sense is coordinating perception, reasoning, and action simultaneously in a program that is limited to language. He also stresses the problems of understanding metaphor, unstated goals, the beliefs of the speaker, and appreciating emotions and underlying motivations in people. Furthermore, there is a central problem of the judgment of plausibility, he says, as in the sentence "Mary jumped ten feet when she heard the news," or, "John ate up the compliments." Understanding plausibility is a step prior to grasping metaphors, humor, exaggerations, or lies. Nor is it common sense, he reminds us, to assume that simple programs can easily be extrapolated to larger ones (133).

The philosophers Hubert Dreyfus and Stuart E. Dreyfus have articulated the arguments against the brain-as-machine position most persuasively. First, they reject the mind-as-machine metaphor and the reason/intuition dichotomy that are taken for granted in technology's metaphysic:

> The truth is that human intelligence can never be replaced with machine intelligence simply because we are not ourselves "thinking machines" in the sense in which that term is commonly understood. . . . We are not proposing to exalt the intuitive at the expense of the analytic abilities so highly developed in our Western culture. . . . The hoary old split between the mystical and the analytic will not do in the computer age. . . . [A]nalysis and intuition work together in the human mind. (Dreyfus and Dreyfus iv)

The brain cannot be simply reduced to a series of on/off switches so that it matches the computer architecture, Hubert Dreyfus argues. Nor can the psyche be reduced to a set of information units without inherent meaning. He also challenges the epistemological assumption made by some roboticists that all knowledge can be formalized into logical relations and rules and be reproduced by machine. He sees this assumption as an unjustifiable generalization from physical science. Nor can we reductively assert that mental operations obviously resemble the processes in computer programs. But Marvin Minksy says: "'That's what the brain is: just a piece of meat that has chemicals and electrical charges'" (qtd. in Dreifus, "Conversation . . . LeDoux"), and, "'Common sense is knowing maybe 30 or 50 million things about the world and having them represented so that when something happens, you can make analogies with others. . . . Emotions are big switches, and there are hundreds of these . . . . If you look at a book about the brain, the brain just looks like switches'" (qtd. in Dreifus, "Conversation . . . Minsky"). Dreyfus counters this claim with Wittgenstein's argument that it is impossible to supply normative rules that prescribe in advance the correct use of a word in all situations (Dreyfus, *Computers Can't*, 71–108). Not just information, but imagination, intuition, and understanding context are essential to human knowledge. "Thus the strong claim that *every form of information* can be processed by a digital computer is misleading" (17). "There are cases in which a native speaker recognizes that a certain linguistic usage is odd and yet is able to understand it—for example, the phrase 'the idea is in the pen.' . . . People often understand each other even when one of the speakers makes a grammatical or semantic mistake" (111). An undetermined portion of unprogrammable, implicit knowledge is required for all understanding.

Two of my favorites examples are my own verbal slips (which my wife never lets me forget): "Never take a dog out of a bone's mouth" and "hordes of hearses," instead of "herds of horses." Would a computer understand my brilliant, original humor and shards of poetry, based on errors? Would artificial intelligence ever understand, "He is so half-assed!" or "I can go him one better"? Language is not used according to strict rules; unconscious, unprogrammable assumptions, creative interpretation, and a sense of humor are frequently needed. How could a computer, massive as its memory may be, store and then choose appropriately from every possible contextual interpretation, if we cannot even predict our own meaningful errors or poetic inspirations? Common-sense understanding is far different than information communication. The context of a situation determines the *significance* or *meaning* of its facts, separating humor from love, for example. Computers can quickly do abstract functions such as scanning texts to find a matching or misspelled word (for which I am very grateful). From a huge number of facts humans can select the currently *relevant* ones and decide their importance. And like other animals, humans can also recognize the meanings of ambiguous patterns in context, such as bared teeth or pulling out a weapon. These subtle, meaning-discerning mental tasks have evaded machine imitation (Dreyfus 128–30, 148–49). Phenomenologists have been saying this for some time.

Another important element of human intelligence is that it is embodied. Pure information in a brain is sometimes imagined as a "brain in a vat" cut off from the body. This grotesque fantasy, removing blood, oxygen, and the entire chemical and nervous system's input, is a naïve illusion. Human intelligence involves a field of relevance, including language and body, that requires a nonprogrammable sense of meaning. "Emotions and concerns *accompany* and *guide* our cognitive behavior," Dreyfus stresses. "Human experience is only intelligible when organized in terms of a situation in which relevance and significance are already given" (288). For example, if one holds up a knife, how would a computer know if it were to be used for a weapon, a meal, or surgery? Computers and robots do *not* think. They never will, because of our endless confusion over the meaning of thinking. How can the thinkers stand back neutrally and observe the extremely complicated and largely invisible thinking process in which they are involved? Very serious beliefs are at stake, Dreyfus concludes: "People have begun to think of themselves as objects able to fit into the inflexible calculations of disembodied machines. . . . Our risk is not the advent of superintelligent computers, but of subintelligent human beings" (Dreyfus 188–200). Would subintelligence include the inability to discern imaginative themes in apparently rational discourse?

The Soulless Machine with Soul

Does a robot have a soul? Is there a "ghost in the machine"? These questions are actually taken seriously by many who fantasize about robots, for the very purpose of the android robot is to imitate specific human capabilities mechanically. For example, in a combat scene in the film *Star Trek: First Contact*, Captain Picard, leading the fight against Borgs that have conquered earth and assimilated humanity into a human/robotic cyborg state, says to Data the robot, who is interested in becoming more human, "Deactivate your emotion chip," as if emotions could be programmed, engineered, and inlaid on a computer chip. The ruthless Borg queen later reactivates the chip.

This mythic theme pictures the crucial argument that human emotions can be downloaded or converted to computer data and treated as such. The same theme appears in *The Matrix*, when humans have to be plugged into the matrix by a phallic rod that is shoved into a tube built into the back of their brains (large enough to destroy their brain stems). The brain has been medically stimulated by electricity for many years, and implantation of electrodes in animal and human brains has been tried since the 1950s. The physician-turned-novelist Michael Crichton wrote about this theme in his 1972 novel *The Terminal Man*. This novel builds on the efforts of neurosurgeons to relieve seizures. His theoretical case is a man with brain damage who blacks out from seizures and becomes violent. He consents to have doctors implant electrodes in his brain, connected to a little nuclear-powered pack under his shoulder skin that relays information to the hospital computer. The treatment goes awry, and he becomes more violent, wrecking the hospital computers, spreading radioactivity, and threatening his doctors' lives. These mythic fantasies and actual experiments reflect the widespread artificial-intelligence argument that human consciousness or soul is "nothing but" digital data that can be easily transferred to computers and switched off and on. The stories are a heady blend of technological extrapolations, dreamy mythologies, and warnings. As Susan Schwartz says: "'Star Trek really *does* want the best of both worlds—the secular humanistic confidence in the triumph of reason, and the emotional and transforming experience of myth'" (qtd. in Kraemer 157). Whether the machine has soul or not is an important component of the Pinocchio Project's search for meaning; its answer would determine an essence of human nature.

The British mathematician Alan Turing created a clever exercise to test this assertion. He asked, "Can computers think?" His "imitation game" was a test to determine whether a computer's operator, out of view, could convince another operator that its messages were human. If so, he argued, then computers could be shown to think. He assumed that "digital computers . . . can in fact mimic the actions of a human computer very closely" and that a child's brain is a "little mechanism" (Turing 438, 456). Thus his premise is included in the conclusion: human minds can be defined as machines. He also reflects the restrictions on questions of meaning: "We are of course supposing for the present that the questions are of the kind to which an answer 'Yes' or 'No' is appropriate, rather than questions such as 'What do you think of Picasso?'" (445). This restriction assumes that simple factual questions can be separated from questions of quality and meaning. This argument is inserted as an assumption in his game. He is necessarily defining consciousness in a way to suit computer languages.

Turing's robot argument would not prove that a functional robot has consciousness, or that humans are nothing but robots, but simply that a mechanistic definition of consciousness, life, or soul can be replicated mechanistically. This would be to argue in a circle of presuppositions. The intellectual trick is that the definition of consciousness, life, or soul must be constructed to fit a machine's requirements. Ellen Ullman calls this fallacy "mistaking the tool for the builder" ("Programming" 65). Many of Turing's arguments begin with the premise that brains are a kind of biological computer, and Minsky glibly sets out his premise: "'We're machines and we think'" (qtd. in Markoff 5). No argument or proof is involved, just the bald assertion of premises, beliefs, and imaginative enchantments. There are reasons for this fascination, of course. When humans learn to speak a computer's language, as when we learn to use a new program, we have to narrow our interaction to fit its specific logic. Computers cannot understand symbolism and nonverbal human communication, such as gestures like the shrug of a shoulder or a wink. These are vague images with many meanings that must be interpreted in context. Of course, computers can perform certain narrow tasks, such as calculation and sorting, with superhuman speed and accuracy, but this is a very limited portion of human consciousness amplified, and it excludes essentially human qualities. Computer on-off electric codes can express only a narrow range of the brain's mysterious totality. It proposes that knowledge is only information and totally ignores the unknown extent of unconscious imagination, intuition, and judgment involved in all intelligence.

It is remarkable that human calculation skills have already been far surpassed by computers that run many of our financial, government, and industrial operations. But if we define the human mind more broadly to include the interpretation of metaphors, contextualization of information, grasping the meaning of a new event, and consideration of ethical issues, then the mechanical definition of mind is too restrictive.

Furthermore, computers are limited and can be misused in ways that trample on human rights. A program that compares faces in a crowd to a database of thousands of people sought by law enforcement is operative, but it is arousing protests against its intrusive violation of freedom. Crowds in a Tampa, Florida, night-life area that is known for crime reacted to cameras scanning them by wearing gas masks and Groucho masks and angrily offering the cameras obscene gestures, saying, "Digitize this!" (Chachere 7A).

Interpreting information requires a human sense of meaning and context that is unprogrammable. Could the face-scanning program by itself understand the symbols, contexts, and ethics of the protesters, agree with them, and decide that its activities are unethical and shut itself down? Not likely, for a police program. As Dreyfus emphasizes, "non-programmable human capacities are involved in all forms of intelligent behavior" (*Computers Can't* 197). Computers lack essential elements of human intelligence such as imagination, meaningful spontaneity, and the general grasp of a situation.

Could a computer fully understand Garrison Keilor's comment that it's "artificial intelligence, but it's better than none at all"? Robots' computers cannot conceive of a different mode of organization than the logic that they possess. A large gap necessarily yawns between mechanistic/electronic machines and human consciousness. Computers cannot come up with ideas that are not previously defined in their programs (Mumford, *Pentagon,* 188–91). Such wishful thinking requires a firm denial or forgetfulness. Computer-driven robots, as Lewis Mumford perceived, have become idols of forgetfulness and deification. Forgetting that human minds invented the hardware and write the software, robot enthusiasts turn their seemingly autonomous inventions into deities, icons of ultimate reality. Then when they dream that *they* are computer-driven robots, they absorb their own illusions of omnipotence (273).

Forgetfulness of the organic depth of consciousness is required by those who dream of the expansive powers of robots. And inflation of the collective ego of industrialism is also required to claim authority for the metaphysic that cannot fulfill cybernetic dreams. The absoluteness of this claim reclaims the authority of religion that mechanistic metaphysics once rejected. Again, the android robot argument is a bold, imaginative enchantment of technology. It is deeply rooted in the imagination, far from pure reason. The more we imagine machines to be like humans, the more we imagine humans to be like machines.

Cosmic Homesickness

Robot mythology is rather primitive, despite its high-tech surface symbolism. As the carrier of the projections of an extravagant trust in and reverence for the mechanical worldview, the robot is an *icon* (Geduld 34). Promoting a powerful, literal reality "out there" in future times, the robot becomes an idol, a postindustrial golden calf. Inspiring obsessive, ritualized devotion, the robot becomes a *fetish*. It is a symbol of the irrational desire to elevate a useful machine and a fruitful intellectual framework into an absolute reality. As Jacques Ellul said, "Technique is the god which brings salvation" (144). Divinity, rejected by the builders of the mechanistic worldview, has come back in a computerized form of the enchantment of technology: Robogod.

Android robot speculation is imbued with a thinly veiled yearning to make robots into godlike beings with absolute power. This yearning reveals an unacknowledged fantasy of their engineers that extends beyond the desire for certainty, for who but gods could make godlike creatures? As Jean Brun says, the desire to create an automaton and give it human capabilities is a dreamy enchantment that is driven by the yearning to grasp godlike powers ourselves, to make humans into gods (Brun, *Masques*, 88–91).

This desire for apotheosis in angelic androids is not only the result of inflated egos; it is a cosmic homesickness, a longing intensified by its denial of the very transcendence that the technological metaphysic denies. As Eric Davis, author of *Techgnosis*, puts it:

> Once we are posthuman cyborgs, all the knobs can be twisted to the demigod settings. . . . [O]ur ability to siphon our minds into any number of possible machines will allow us to explore deep space, colonize other planets, and mine wealth from the raw stuff of the solar system. . . . [This is] the heroic, otherworldly dream to leave the planet, a dream that sums up the transcendental materialism . . . [of] technophilic mutants. . . . They also literalize the cosmic homesickness that vibrates in so many human hearts, a longing for a transcendental level of authenticity, vision, and being reflected in the heavens. (127)

We certainly should use and welcome the work-saving powers that computers bring and celebrate the remarkable skills of our creative robot engineers. But their speculations need to be seen as speculations and not literalized into naïve beliefs, and their theories need to be subjected to philosophical, psychological, and spiritual analyses to avoid destructive distortions of human nature. For example, I would suggest that we dispense with the term "artificial intelligence," since it incorporates only a fraction of intelligence, and replace it with the term "machine logic." The notion of artificial intelligence is a dried-up skeleton of a theory of mind that depends on a magical leap of faith to believe that the dead bones will in some inexplicable way give birth to all the organic, embodied foundations of the real living mind. Selected logical functions can be programmed to exceed selected aspects of the mind, but the full context of the mind—especially its enchantments—cannot be programmed.

To imagine that machines have intelligence and think is a naïve anthropomorphic projection of human qualities. We should stop using these playfully poetic but misleading words. It is a sign of a hypertrophy of the mechanical mind. The nicest people say the strangest, even most contradictory, illogical things, such as: Computers do not have emotions, only parallels to emotions, but we program "emotions" into them. They say that aspects of human consciousness such as empathy are mechanisms, that computers have attention, perception, social interaction, intentions, desires, anger, fear, sadness, surprise, and beliefs. They say that computers have eyes, mouths, ears, lips, eyebrows, and so forth (Breazeal).

But these are simply the artificial simulacra of human minds and organs. Robots do not have eyes; they have cameras. They do not have ears; they have microphones. Robots do not have intentions or desires; they have computer programs. Now their robot engineers know this, of course, but they suspend their disbelief and speak poetically about eyes and ears, emotions and social interaction, treating their androids much as they would a high-tech puppet or doll, trying to play Pinocchio and bring it to life. These huge category mistakes leap from electronics and mechanisms to emotion with no evidence, no argument, no more than a hopeful skip and a hop. Poetically calling a camera an "eye" does not make it literally an eye, or even a camera on the way to becoming an eye.

We need to practice vigilant observation of our theories and visions. We need them for inspiration, but we must honestly acknowledge our imaginative desires and assumptions, in case they attempt to make end runs around logical reasoning with crude a priori claims to absolute truth. We will not eliminate the imaginative enchantments of robot technology, but we can take responsibility for them and, in harmony with the best of reason, bring them to a high level of cultural refinement.

The spiritual and the technological inevitably work together, and they belong together. We must be conscious of intuition, imagination, symbolic interpretation, and contextual analysis as well as mathematical measurement, logic, and engineering design, if we are to avoid computer illiteracy, symbolic illiteracy, and exaggerated claims to absolute knowledge. Our concept of reason needs to expand so that it acknowledges not only logic, quantity, and technical excellence but the finest qualities of life and caring, such as human imagination, interpretation, sense of purpose, morality, forgiveness, compassion for the other, and a sense of social justice. The modern superficial, conscious construct of reason lacks these qualities, but they can be unearthed intuitively because they unconsciously already pervade the supposedly purely rational arguments of real humans.

8   Being Enchanted

No one has ever been modern. Modernity has never begun.
There has never been a modern world.
    —Bruno Latour, *We Have Never Been Modern*

Big Success, Big Failure

The painful paradox of modern technology is that it has succeeded wildly, yet it has failed miserably. We live in this paradox that questions the very meaning of being modern. Scientific technology has succeeded at making amazing machines, from power tools to rockets, at improving health and increasing life span, at producing better food, shelter, and an abundance of work-reducing life comforts. All these benefits depend, of course, upon your nationality, social class, and race, but they are rapidly spreading around the world. Modernism has brought many individual freedoms and human rights, for which we can also be grateful. No longer are people tied to farms and backbreaking labor. Many can get an education, travel, communicate widely, share governance, change careers, and control birth rates. These successes are obviously valuable.

But, as I have argued throughout this book, despite all our high hopes, technology has also sadly failed because of our enchantments, which have allowed its utopian projects to be blindsided by intentional misuse, extravagant, utopian dreams, and "unanticipated" side-effects. As systematic processes and as concrete products, technological culture has failed because our enchantments have led many to see technology as an unstoppable juggernaut, an autonomous, disenchanted, objective realm with no adequate human direction. Industrial society has failed because we have let it proliferate full of reckless unconscious dreams, creating serious disasters, such as nuclear hazards, and breeding blatant injustices by creating many conveniences for the rich and ignoring the poor of the world. We are rapidly approaching an unsustainable technological culture that is already affecting the entire world, in global warming and extinctions of species. Technological society has also failed because it is so enchanted by its dualistic ontology that it promotes the painful nihilism behind the flaws and blindly covers up the unified depth where its potential healing could take place.

Industrial society seems triumphant because it has successfully created powers and pleasures for a minority of the earth's population by exploiting knowledge of a fraction of the whole range of truth and rapidly overusing valuable resources from around the world. But its success is built on the narrow, materialistic subject/object metaphysic that often claims to be able to explain the full range of reality. Its advocates interpret this dualism as an unquestioned premise rather than an optional choice. Technological culture is so enchanted with technology's powers that it forcefully suppresses, marginalizes, and scorns what it cannot explain by calling it "subjective," "illusory," "romantic," "flaky," "fuzzy," or "unscientific." This way of maintaining a worldview borders on obsessive devotion to a one-sided mental style rather than truly objective logic. But much of the discarded realm includes important ethical and spiritual realities that it cannot measure, but which it needs to guide it out of its increasing ecological, ethical, and nihilistic crisis.

This imaginative dualistic worldview gives meaning to its devoted followers, offering concrete lifestyle improvements, power, pleasure, wars, and even a quest for transcendent significance in space travel and science fiction. While the finest benefits of modern technology should be maintained, the failures will outweigh the successes if the flaws of the technological system are not quickly corrected. The meaning supplied by technoculture's worldview is inadequate spiritually and ethically, for it promotes a dangerously narrow dualistic metaphysic, excessively crude values, poor self-awareness, and shallow ethics. A great awakening from these enchantments is urgently needed.

The Enchantment of Disenchantment

The greatest enchantment of the industrial world is the belief that it has disenchanted the world, denying the soul and Being in order to inflate the power and domination of humans in a world of objects. As we have seen with speed, utopian triumphalism, the space cowboy, and the robot, technologies in the lifeworld that are outwardly thought of as disenchantments are full of enchantments. But the denial of this is the first enchantment. Although legions of scientists and engineers have the highest standards of neutrality and fairness, others promote naïve materialism, and at some point in the production, sales, or application process, this is too often lost or denied. Such denials are far from objective, rational decisions; they are the products of massive nihilistic enchantments, power- and profit-driven fictions that technology is neutral. As the sociologist Stanley Aronowitz says, "It does not matter that the scientific community ritualistically denies its alliance with economic/industrial and military power. The evidence is overwhelming that such is the case" (20).

A very important line of discussion of this problem is the argument that since the world has been disenchanted, it needs to be reenchanted in a new way. An excellent statement of this argument is Morris Berman's *The Reenchantment of the World*. Berman lays out well the process of disenchantment in the development of scientific consciousness and the current signs of the dissolution of the strength of that metaphysic. His overall thesis differs a bit from mine, though: I am proposing that the process of disenchantment has not been nearly as complete as Berman assumes. I argue that enchantment has always remained an effective but denied factor in technological thinking, just beneath rational consciousness.

Yet these two approaches overlap, because Berman says, "The sickness of our time is not the absence of participation but the stubborn denial that it exists" (*Reenchantment* 180). This aspect of his analysis is what industrial society has neglected, the fact that knowledge and technology are never fully disenchanted but, just under the surface, are always passionate desires hard at work guiding and shaping technologies.

To assume total disenchantment is to accept only conscious thought as significant. But when we allow consciousness to include unconscious input, we can see the barely concealed enchantment always at work. So the process of reenchantment can best be undertaken by showing not only that the world is disenchanted and we have to import a new consciousness but that the enchantment never left. The modern subject/object dualism operates only at a surface level. As the economist Herbert Simon says, "Science is inextricably interwoven with cosmology, with theology, with epistemology, with ethics, politics and psychology" (247). Although at a conscious level we think that science and technology have dismissed more humanistic and spiritual factors to the sidelines, these factors have always been fully involved with technology under the surface of consciousness. They are an essential aspect of technological development, but, lacking adequate consciousness, they easily remain crude. When these concerns are denied in debates regarding the role of technology in culture, then unacknowledged and thus more primitive motives guide technology. Suppression of refined ethics gives free rein to the raw will to power. As Roy Helton says,

> [M]achinery expresses only one trait in man, and that one trait is his craving for power. Power is its achievement. Power is what it is for—power over nature,

> power over people, power over time, and power over space. Now power is good, but so is love good, and so is beauty good, and so is quiet good, and so is truth good, but none of them is adequate alone as a basis for civilization. . . . Machinery has not drained off and sublimated man's hunger for power. It has fed that hunger, and lent it teeth and claws, and also has sanctified it. (143)

Modern technology has increased the mechanistic power under human control, and the suppression of ethical guidance has multiplied its effects immensely—from the dangers of nuclear power to those of ocean overfishing. Our enchantment with technology has fostered this denial. What we need is not only a reenchantment but a deeper reawakening of ongoing enchantments—some crude and nihilistic, some magnificent—so we can take responsibility for them.

Berman rightly argues that the reawakening of enchantment is first of all a reembodiment of knowledge, a shift out of the isolated ego standing against and dominating the world to a fully embodied consciousness:

> What I mean by "Mind" is the conjunction of the world and the body—all of the body, brain and ego functions included. . . . Only a disembodied intellect can confront "matter," "data," or "phenomena"—loaded terms that Western culture uses to maintain the subject/object distinction. (*Reenchantment* 183)

Like the Headless Horseman, the disembodied subject must rejoin reason with the entire environment of body, soul and nature, imagination, intuition, ethics, and spirituality. Building on Reich, Jung, and Bateson, Berman sketches a new ontology in which we must be able to stand outside the entire technoculture field and see its contradictions as if it were a schizophrenic system (228–29), because, "By and large, it is a charade. The modern reality-system requires allegiance to a logic that in actual practice has to be violated all the time. Western society has deutero-learned [i.e. learned second-hand] a Cartesian double bind and called it 'reality'" (232). In a new ontology, "Fact and value are not split, nor are 'inner' and 'outer' separate realities. Quality is the issue, not quantity, and most phenomena are, at least in a special sense, alive" (237). In *Coming to Our Senses*, Berman elaborates: "No matter how 'spiritual' the cosmological urge may seem, it is, I want to emphasize, thoroughly grounded in the tissues of the body; meaning is part of the bodily self" (43).

According to Berman, participation in the wholeness of the world is needed, but the biblical traditions will not support that: "[T]he Judeo-Christian tradition sees us as masters of the household, Batesonian holism sees us as guests in nature's nest" (*Reenchantment* 259). Berman rightly sees the need for meditative experiences in an embodied spirituality (286–87). Of course, the meditative feeling of a cosmic connectedness gives meaning. But for Berman there is no God, or no ontological ground for holistic consciousness. "What would be worshipped, if anything, is ourselves, each other, and *this earth—our home*, the body of us all that makes our lives possible" (280). This stops short of the essential ontological question of the ground of existence. It may be appropriate for an anthropological method, but it is necessary to confront ontology for a complete consideration of the problem. At this ontological level we can question how we have lived in modernity's proposed subject/object dichotomy. Awakening to enchantment will require more self-awareness and further awakening to the ontological depths of existence itself.

We Have Never Been Modern

As we ride the wave of technological culture's evolving destiny, with its positive and negative outcomes and its constant breaching of its own dichotomies, we are pressed to ask a fundamental question: Have we really ever been modern? Have we really "grown mechanical in head and heart, as well as in hand," and is "our true Deity Mechanism," as Thomas Carlyle proclaimed in 1829 (63)? Is modern numbness to soul complete? No. Many may think that way, but we don't all act that way. Think about Robot Wars' primitive technoviolence and flaming nihilism. This drives their tinkering, not reason. We have never been totally objective or purely subjective, so we have never been totally modern.

Of the many thinkers who have criticized modernity, one has pressed the issue exactly this way. Bruno Latour, a sociologist in Paris, entitled a 1991 book *We Have Never Been Modern*. Latour argues that the negative consequences of technology have taken the wind out of the sails of the industrial project. Faith in modernity has lost its conviction. Its challenge of emancipation has been stifled, and its goal of domination has turned sour (Latour 9–10). In this view, the protests against technodomination and the talk of postmodernism are windows to the future, and the fierce political/corporate defenses of industrialism's profits are a backward-looking rear guard. The history of modernism was not so progressive as we thought it was.

Latour calls modernity's theoretically distinct ontological zones—the subjective human realm and the objective arena of nonhuman thingness—the Great Divide. The gap is theoretically crossed by Kant's a priori categories of reason, such as time, space, and causality. It is also crossed by emotional, subjective "projections," as Freud called them, which theoretically cloak the objective world with fantasies such as love and fear, and which should be "withdrawn" back into the subjective containers to reestablish objective "reality." These mental operations are classic modern tools incorporated into the basic metaphysics of the Great Divide that help explain its incompleteness. This is the modern "Constitution" that "invents a separation between the scientific power charged with representing things and the political power charged with representing subjects" (Latour 29). But there is far more in between than we perceive.

Relativity and quantum physics point to the unity. They have shown that those "objects" reduced to atoms "out there" are not really solid things but rather fields of energy. Things, or "objects," are masses of dancing energy fields. Mind, or "subjectivity," is also an energetic field that is not simply contained in one's brain. The moon is not just a thing floating in empty space but is *part* of the electromagnetic field of the earth. The earth's electromagnetic field is like a giant magnet, directing our compasses and forming an invisible shield against immense and deadly solar radiation storms. It also forms the subatomic structure of our world and bodies. We cannot stand outside it and observe electromagnetism. Nor is it only internal, so it is neither objective nor subjective; it shapes the very structure of existence, from the subatomic components of life's cells to cosmic gravitation. This basic energy is closer to the ontological ground of Being.

On the human scale, Latour says that there is a vast realm between our postulated subjects and objects that he calls hybrids, mediators, quasi-objects, or quasi-subjects: "[U]nderneath the opposition between objects and subjects, there is the whirlwind of the mediators" (46). Here we have the phenomena we have been discussing, such as the sublimely streamlined vehicles of speed; the utopian fantasies that become sinking ships; the space cowboys and their expensive, dangerous rituals of frontier weaponization; the robots with their iconic quest for absolute mechanization and destructive "gearhead" nihilism. These are all hybrids, operating de facto between the theoretical subject/object divide. But the technological Constitution forbids the conceptualization of the many hybrids between the ontological Great Divide:

> [H]ere, on the left, are things themselves; there, on the right, is the free society of speaking, thinking subjects, values, and of signs. Everything happens in the middle, everything passes between the two, everything happens by way of mediation, translation, and networks, but this space does not exist, it has no place. It is that unthinkable, the unconscious of the moderns. (Latour 37)

The more we examine the unthinkable intermediate realm where opposites coincide, the more the forbidden conceptualization dissolves. When a theoretical dichotomy dissolves into such a paradox, it is time to question the supposed duality's validity.

Latour argues that "the modern world has never happened, in the sense that it has never functioned according to the rules of its official Constitution alone" (39). Even quintessentially "modern" traffic jams and skyscrapers are hybrids, products of irrational, unecological market drives and the will to speed and power as well as concrete and city planning. They are unthinkable, theoretically deniable phenomena. Such hybrids are concretized pride and aggression in steel, glass, and trucks.

The moderns think that they have succeeded because they have effectively thought objectively, ruling out the subjective. On the contrary, they have constantly blended the two, repeatedly denying the subjective while letting it have free play in the objective world, which means that the separation has not happened. It wasn't blended because it wasn't ever separated. We just thought of it as dual, so we have never really been modern. We have become highly technological but not objective. Our machines and their environments embody desire as much as reason.

Latour disagrees with the charge that modernity has forgotten Being—the fullness of existence itself—because he thinks this is impossible. "Who has forgotten Being? No one, no one ever has, otherwise Nature would be truly available as pure 'stock'" (66). Networks and machines are full of Being, for it is impossible for anyone to forget Being. This is an interesting question at the heart of Latour's view that we have never been modern: Is it impossible to forget Being and totally identify with modernity's subject-object Constitution?

No one has the power to fully forget Being, I would argue, but humans have the powerful ability to repress Being, to create mental worlds, deny anomalies, and convince themselves of the totality of a dualistic metaphysical construct. So repressed Being *seems* forgotten, making room for advanced technology and nihilism. But under the surface, unconsciously, remembrance of Being remains, even if it is wrongly shriveled to merely subjective feelings of love or anger, or aesthetic appreciation of a national park or art.

Modernity was not a period that overwhelmed all previous eras but a hybrid of old and new, of ageless archetypal desires and new inventions. Modern inventions can embody and serve the most archaic purposes—violent and crude, or elegant and caring. Some modern skyscrapers house businesses whose archaic profit drive helps promote the ancient poverty and crime around them. No "modern" object exists without its broader context in vast Being. Latour says that we are all "polytemporal," mixing up transistors and tattoos, plastic and animal skins. Thus "modernization has never occurred" (75–76).

We have sought to apply the reductionist method of eliminating "extraneous" phenomena and narrowing consideration down to the objectivist methodology of science to just about every cultural phenomenon, from atoms to God. But it only works to the extent that we agree to eliminate the "extraneous" factors, like the business that attempts to shove off its "external" costs, such as pollution, to society. Once we see through the method, we can admit the other factors into a holistic view. Latour says, "Reductionism has never been applied to the modern world, whereas it was supposed to have been applied to everything!" (116). Although laboratory tests proceed by reduction to constants and variables, the choice of constants and variables—the measuring system and instruments, the purpose of the experiment, and the applications—all contain nonobjective factors that cannot be eliminated, so total reduction is not possible. Modernism's program imagines itself possible only by neglect.

How shall we evaluate the technologies of modernity? Obviously, we need scientific method and lab research, but we must also acknowledge their broader context. The effort to achieve a neutral, detached stance protects and enlarges knowledge, but to remain detached about the broader context is dangerous. About this refinement of consciousness Latour says:

> The moderns' greatness stems from their proliferation of hybrids. . . . Their daring, their research, their innovativeness, their action, the creation of stabilized objects independent of society, the freedom of a society liberated from objects—all these are features we want to keep. On the other hand, we cannot retain the illusions (whether they deem it positive or negative) that moderns have about themselves and want to generalize to everyone: atheist, materialist, spiritualist, theist, rational, effective, objective, universal, critical. . . . Let us keep what is best about them. (133)

The difficult task of deciding just what is the "best" of modern technologies is made easier when the desires and enchantments of the technical systems are openly acknowledged. Some enchantments are dangerous, such as nuclear power and weapons (that may leak or be stolen); some are quests for pleasure, such as saline breast implants (that may leak); and some are just expensive, such as the Concorde, a beautiful French supersonic airplane that offered trans-Atlantic flights at twice the speed of sound for three decades. However, this great engineering feat finally failed because it was so expensive. It burned twice as much fuel as a Boeing 747 and carried only a quarter of the number of passengers. For several thousand dollars, the rich could sip champagne while seeing the earth's curvature. But finally, it did not pay for itself. After a tragic flaming crash in Paris in 2000 and the need to replace aging planes, the Concorde was retired, along with plans for Boeing's similar Sonic Cruiser. The failure of a plane of the future to work amid economic realities has dashed one more utopian dream ("Farewell").

Finally, Latour's vision frees the world of disenchantment. If we recognize the artificiality and narrowed view of subject/object metaphysics, we see that it is an ideal conceptual typology that has never been fully enacted. Then we can begin to reclaim the glimpses of Being that constantly peek through the facade of modernity—the love of nature, the social community that machines have too often replaced, the sense of the awe of the universe when we look up at the stars, and the feeling of being at home on this planet. The harshness of objectivity's one-sided consciousness has suppressed these realities for too long. As Latour says, "Haven't we shed enough tears over the disenchantment of the world? Haven't we frightened ourselves enough with the poor European who is thrust into a cold soulless cosmos, wandering on an inert planet in a world devoid of meaning?" (115). The modern vision of a disenchanted, detached, authoritative science and technology is no longer possible.

Drinking from the Deepest Well

The theoretically separate entities of object and subject have become conventional intellectual tools stretched beyond their own realms to include the many phenomena of the in-between hybrids, as we have been showing in the enchantments of technology. This discussion will require a poetic language that points to deep realities. Instead of constantly trying to squeeze phenomena such as robots back into the dualistic framework, we might as well let go and accept our participation and source in the deep, prior well embracing opposites, the well older than the cycling planets and the sparkling galaxies.

The blend of desire and reason in technology is a union of opposites prior to dualistic, conscious thought. It is not a state of confusion but a deep, ontologically prior zone of existence, going all the way down to the source, constantly giving birth to all existence. Linguistic discussion of this invisible realm is difficult, because it preexists human logic and conscious reasoning but is simultaneously the root of consciousness. In philosophy and religion, this zone of existence is the deepest glimpse of ultimate reality, the ontological source of all existence that humans can achieve, called by many names: God, Being, Emptiness, Nature, or Ultimate Reality. But we are like mere ants crawling up a tree trying to explain the tree's system of existence, from seed to branching life form dependent on soil, air, light, and water. Despite our bold claims for the scope of reason, any discussion of this primordial mystery must include images and concepts from earthly experience. No experience, no language can adequately express its timeless, deathless fullness without reference to experiences in time/space.

Thus, the language with which we mere morals attempt to express our glimpses of the *unus mundus* has different patterns than ordinary language. This realm is not a thing, for it is the origin of all things in the universe, so the language of thingness is unsuitable to express it. Ordinary language will never plumb the bottomless well and, due to the vastness of the waters, ordinariness will be denied the certainty that it seeks in rational, logical metaphysical discourse. Yet the waters of the well infuse existence, offering us endless drinks of its fruitful energy.

For example, conventional religious language calls this reality "God," the creator of the universe, and proceeds to endow it with endless metaphors, such as God the Father or Mother Goddess. However, those who seek to guarantee its certainty misread these anthropomorphic metaphors as literal statements. For example, how could the source of all existence literally be restricted to one half of a reproductive pair on one if its tiny planets?

Philosophers and theologians who seek to avoid religious metaphors use abstract terms such as "Being." Parmenides and Plato began the western tradition of the philosophy of Being, which is full of difficulties, such as whether it is an unchangeable realm above the world of becoming, and if so, how they are related. Alternatively, if it pervades existence, how is it related to changing and dying things? This approach helps clarify the issues conceptually but suffers from the limits of abstraction and logic, whose structures are not guaranteed to touch the eternal ground.

A third approach acknowledges that neither literalized earthly metaphors nor clear logical distinctions about the unifying ground of existence are adequate. Its grand mystery is too majestic and transcendent for us to fully grasp. However, our relationship to it is too important to ignore, if we are to transcend nihilism. We are gracefully granted glimpses of its wonder beyond every experience. Both transcendent and immanent, creator and holy spirit, the first world is also an eternally present Being-in-the-world. Awareness of this will erase reductive enchantments about robots becoming human and other enchantments of technology.

The End of the Subject

The subjective ego, as part of the metaphysic of industrialism, promotes exaggerated self-importance, isolation from soul and from nature. While denying or neglecting this isolation, ego builds massive technological frameworks based on the desire for domination. Its ultimate enchantment is its inflation of itself to godlike status. It takes for granted vast powers such as speed, flight, and destructiveness, once reserved only for gods. But the subjective ego is only a cultural construct, an ideal that in fact never contained and cannot control all the experiences that the moderns attempted to squeeze into it. *We should drop the concept of the subject and redefine it as "soul,"* as James Hillman and others have done in archetypal psychology. Soul includes a more humble ego, a more accessible unconscious psyche, and a more natural, connected role in the world.

Hillman has explored paths of imagination beyond the Cartesian subject/object dichotomy, and he sees a more primeval soul-in-the-world. Building on Jung's view of the collective unconscious, Hillman stresses that we are primarily image makers, and what we think of as subjective appears in images, both interior and exterior:

> We live in a world that is neither "inner" nor "outer." Rather the psychic world is an imaginal world, just as image is psyche. Paradoxically, at the same time these images are in us and we live in the midst of them. (*Re-visioning* 23)

A technology of soul would be guided by honest feelings, serious ethics, elegant aesthetics, and ecological balance rather than ego's lusts for power, domination, and pleasure, thinly veiled by curious ideas of "reason." We would see the end of spirituality as "innerness" and begin to feel our immersion in Being-in-the-world, in the ancient atmosphere, the waters, forests, mountains, gravitational fields, and stars. We would begin to feel that we *are* star-energy, body and soul, blessed with sharing together a few years on a beautiful planet, to learn, and to care.

Soul-in-the-World

Whereas the industrial ego imagines itself in control, represses its unconscious, and forges ahead with unconsciously shaped technological efforts to dominate the world and other people, soul-in-the-world feels its deep connection to things. Soul-in-the-world is so contrary to objectivity that it is feared and suppressed. But opening to it is rewarding. When we open to soul-in-the-world, we can savor the comfort of home in a house, just as a fox has its den. We can admire the natural power of spots such as the Grand Canyon, which makes us feel so small, yet gifted to share in its grandeur. When we walk along the seashore we can feel the delight of ageless sands under our feet, the splashing of the ancient tides pulled by the moon's gravity, and relish being at the border of a vast body of water extending, in huge waves, to other continents. We can remember our dependence on water for life and that the ocean makes life on this planet possible. These places are not simply objects in an alien world. This is our home; we belong together, and this deep belonging is soul-in-the-world. We fit, we participate in soul-in-the-world like a nest in the cosmos. In *The Practice of Technology,* Alan Drengson argues:

> [T]he small personal self (as ego) is embedded in a larger, ecological, transpersonal Self. . . . The "separate" ego sees the world in terms of friends, enemies, and heroic struggles. The larger Self *knows* that the world is filled with possibilities for *reciprocal, complementary,* conscious relationships. (48)

A technology of soul-in-the-world asks, "Whom does this technology serve?" and is wary of ego's answers. Ego can build ugly houses, tolerate poverty-ridden regions, or create towering glass and steel skyscrapers with windowless cubicles that express alienating, domineering power more than a caring habitation for work. We project our own personal and collective shadows in cruel, stereotypical prejudices. But when we become conscious of soul-in-the-world and take ethical responsibility for it, we can make qualitative judgments and choose refined, elegant, and ethical soul in culture and nature. When in harmony with the larger soul, the smaller conscious ego does not dominate but *serves* the higher principles of the Soul, Self, or Being-in-the-world. Symbolized in many forms—mandala, sun, crystal, or light—soul sheds ego's anxiety, denial, and isolation, embracing the larger context of community and openness to intuitive knowledge.

Soul still creates and uses technology, but it is guided by different principles than the ego. Soul's technologies derive from openness to Being. The conflict between these approaches will continue to be a hot political issue. Industrialism's collective ego wants to extend the modern project of exploiting and dominating nature, extracting and consuming resources as rapidly as possible for the maximum profit of a few. Ecological disasters are ignored or minimized and pollution grows while resources diminish. By contrast, soul-in-the-world uses technologies of conservation of resources, such as minimizing resource extraction, maximizing recycling and reusing, and minimizing consumption. Ecological soul-in-the-world embraces the wilderness, equal distribution of wealth, and meditative consciousness because it feels empathy for the world and is rooted in the care of Being, not domination.

The End of Innerness

Even some very soulful people are still forced by our linguistic conventions to refer to religion as "innerness." But this word leaves the "subjective" head cut off the horseman, who still rages out there in search of more power in objectivity. While spirituality is often present in quiet, private feelings, Being is also present in the depths of the world outside, in a circle of friends, a love, or the moonlight. Spirituality must reclaim its place not only inside the head but prior to the division of in/out, mind/body, and head/horseman. While conscious distinctions between inside and outside are a practical convenience in the ordinary world, technological philosophy has enchanted us with the fantasy that the subjective/objective world is all there is. However, when we see the soul of the world "out there" as well as "in here," that will be the end of innerness and outerness as resting places for reality.

James Hillman says that "[t]he sense of 'in-ness' refers neither to location nor to physical containment. It is not a spatial idea, but an imaginal metaphor for the soul's nonvisible and nonliteral inherence, the imaginal quality within all events" (*Re-visioning* 173). As a psychologist focusing on imagination, Hillman stops short of ontological reflections and refuses to allegorize dreamy images into reified conscious constructs. He sticks with the images and drags the ego down into the dank basement of the soul, rather than pulling the depths up for abstract categorization. Most important, he shifts our gaze from interiority to soul-in-the-world. His colleague Thomas Moore agrees:

> With a modern twist, Hillman revives the ancient notion of *anima mundi*, the soul of the world. In this view the individual as well as all things are perceived to be ensouled or in soul; soul is not to be found only within the individual. (Moore, "James Hillman," 279)

The world will be deeper when we are fully embodied, embraced in the environment, standing outside the subjective/objective metaphysic. The stars will be older, the mountains will be grander, the seasons will be more wondrous, the weather will not be there only to please me, and life will be a glorious gift from the enduring presence of Being.

When we release the image of the self-sufficient, sovereign, autonomous subject controlling the body as if it were other, we flow into embodiment. The higher thoughts of consciousness depend on the support of the unconscious, the heart, community, food, air, water, planet, sun . . . Being. We can cut a flower and enjoy its beauty, but without its water, sun, roots, and plant matrix, it will die. Similarly, we can cut the subject off, but when we reify it as a separate entity, it suffers by losing contact with deeper soul.

The end of innerness as subjectivity is not the end of responsibility, self-control, and independence. But it is the end of humanism, which sees itself as guiding our fate with no reference to any greater Being. Thankfully, humanism has freed us from many false authorities, but it remains stuck in the absolute faith in human reason's dominion of nature. Its values have failed to provide a strong alternative to the past that it replaced, and it has not prevented modern nihilism.

Returning to the fullness of Being by no means eliminates privacy, a personal life, reflection on the self, and quiet, introverted times. Indeed, these may be the richest moments of the presence of Being. But there is no escape in contemplation, for it calls us back into the active life, to life fully lived out of more depth. It calls us to more awareness of our archetypal roles as friends, lovers, parents, workers, and socially responsible world citizens. It calls us to care. At the end of innerness is a memory of the ancient Greek admonition carved on the temple at Delphi and quoted by Socrates: "Know thyself." Like the ancients, but in a new way, incorporating but going beyond science and technology, knowing oneself can now become a knowing our soul-in-the-world again, because we have always been in it.

The End of the Object

The obtuse-object enchantment is a powerful captivation: the notion that the world is composed mostly of soulless, purely material objects "out there" that must be literally interpreted through rational ego-consciousness. This enchantment distorts the world by thinking that the conscious subject is in control of knowledge and certain truth. Numerous experiences that challenge this charm are denied in the name of maintaining the enchantment. We have seen that the *Titanic* was not simply an object, that spaceships are not simply objects, and that robots are not simply objects.

The "things" of this planet need to have more of the fullness of their Being made conscious. A table, for example, is not only a factual, technological object but a part of the social construct of sitting in a chair and focusing consciousness on a flat surface, rather than sitting on a cushion and eating food on the floor, as is done in some societies. A table is a part of the nature of the wood or other materials from which it was made and the earth's ecology that produces its raw materials—the tree, its species, its time and place as it grew, the water, soil, and sunlight that nourished it. A table is a particular example of the design of its designer, part of the historical period it represents, and part of the care that it receives in maintenance. A table may be part of a carpenter's rugged workshop, or part of an elegant meal, central in a family's soul. It is not an isolated thing but a participant in these soul phenomena. No table is an isolated object, a fact free from its part of the continuum of soul and Being-in-the-world.

If the things of the world are not mere materialistic objects, how shall we speak of them? I suggest that we simply use the word *creations*. The world is full of the particulars of Being—hills as old as the stars and parts of human cultures, from alphabets to rockets. *To speak of "creations" instead of "objects"* should connect us back to the whole continuum instead of disconnecting them to make them available for exploitation. We should not imagine creations as the works of an old-fashioned theistic, biblical, anthropomorphic God. Creations can be seen as the products of more abstract Being. The word "creation" can connote *participation* in Being-in-the-world's holistic ontology rather than thinking objects for subjective ego-control. Every creation—plant, insect, machine, person—participates in Being, which makes its very existence possible. Humans are creations, and we make creations in art and technology, but the materials we use are creations of the larger universe—stone, wood, iron, silicon, and oil. Using the word "creation" instead of "object" leads thinking back to this larger context and promotes respect.

This does not mean the abandonment of scientific thinking and technological development. We should by all means take advantage of the brilliant accomplishments of the best of scientific methodology and technological practices, free of the enchantment of objectivity. But remembering the underpinning ontology broadens the horizons of hypotheses and experiments toward a horizon not of domination but of participation in Being-in-the-world. Instead of objects we can look at creations and facts about them not only as instances of scientific categories but as particulars in the vastly larger depth of Being-in-the-world and soul-in-the-world.

It Hath No Bottom

On a deeper level than as culturally constructed subjects and objects, we participate in a largely unconscious reality that hath no bottom. The term *unconscious* is inadequate for this realm has long been known by many other names. It is not only comprised of consciousness, desires, and dreams but of Being, of bottomless ontological reality. In Shakespeare's *Midsummer Night's Dream,* it echoes in Bottom's dream:

> I have had a most rare vision. I have had a dream—past the wit of man to say what dream it was. . . . The eye of man hath not heard, the ear of man hath not seen, man's hand is not able to taste, his tongue to conceive, not his ear to report, what my dream was. . . . [I]t shall be called Bottom's dream, because it hath no bottom. (4.1.5–18)

For millennia, large portions of culture lingered in unconsciousness, expressed in bottomless myth, ritual, and *participation mystique,* or the experience of participation in Being (Plato, "Gorgias," "Phaedo"; Levy-Bruhl, *Primitive* and *Notebooks;* Skolimowski). Creations are revealed by scientific knowledge, but they also have deep roots in unconscious soul-in-the-world. An airplane, for example, can be felt as an expression of the yearning for transcendence, not just speed and power.

Ego-consciousness also has deep roots, especially in its quest for the certainty of its world-construct. Ego hears unconscious heroic archetypal depths seeking control, but also others that challenge its certainty. This adds to ego's anxiety, which produces rigid, restrictive defenses. As the philosopher David Michael Levin says,

The objectification which re-presents the visible in its absolute otherness is, however, a reflection of the *ego-logical need for security*—a need which reappears in the madness of metaphysics as "the quest for certainty." The ego is always—and of necessity—attached to the issue of certitude. (66)

Freud, Levin reminds us, showed that the ego is the product of anxiety that claims to be able to determine a certain external reality. But the rigidity of its system intensifies the anxiety and thus the defenses—which, Levin rightly says, are externalized in global militarism—that amplify ego's basic anxiety in technological form. Ego's anxiety participates in weapons, not as objects but as embodiments of ego's anxious, frightened, even paranoid defenses, dangerously out of touch with the depths of soul from which defenses always come. Ego imagines that it reaches the bottom of the soul's bucket, but it must keep repairing its version of the bottom of reality with endless defenses and their technological expressions guarding its imagined certainty. Huge weapons are ego's fearful anxiety built in steel and explosives (89).

By contrast, the bottomlessness of the pregnant cosmos is gracefully expressed in the astrophysicists' awesome photographs of deep space—its vast galaxies and colorful clouds of the universe's raw materials. The inconceivably immense size of the universe beyond our grasp evokes a humble wonder. But ego's dominant motive for space exploration is the wish to know our cosmic origin, so that we might gain a godlike view of the depths of the primordial. Science-fiction adventures into outer space reflect this enchantment. Images of spaceships zooming past stars at impossibly fast "hyper" speeds toward the endless reaches of the unknown expresses the industrial ego's fascination with perpetual conquest.

But this journey also discloses a thinly veiled spiritual quest for meaning, the meanings of Being neglected by those souls lost in space. The bottomless, endless depths of Being are pictured as empty of care by egos seeking meanings, but they cannot see it right in their face because their defensive egos are enchanted by domination. Below this drama is the lure of the vast reaches of the stunning, humbling mysteries of the timeless beyond. The delight at discovering the familiar structure of our scientific knowledge out there in planetary clouds of sulfuric acid or methane is swallowed by the revelation of the universe's absolutely awesome depth of soul. The robotic techniques of objectness get us out there, only to discover our tiny smallness in the overarching wonder of creation. Have these creations existed without human participation? Of course, but not without the bottomless well of Being.

Being-in-the-World

The strongest critical analysis of subjectivism comes from phenomenologists who reject the subject/object dualism and explore in various ways the immediate experience of the lifeworld (*lebenswelt*). Edmund Husserl, the founder of phenomenology, stressed the prior reality of the lifeworld, in which all humans participate, that sees any subject/object dualism as a convenient construct, not an absolute, metaphysical dualism. He argued that science is a human spiritual accomplishment that presupposes an intuitive surrounding world of life for all. Even the physicist, making careful quantitative measurements, must be considered to be involved as a person in all of physics' theoretical ideas ("Crisis" 121). Martin Heidegger expanded the phenomenological method to focus on ontology—the study of Being. He wrote *Being and Time* in 1927, which astonished the German philosophical world. In it he opens anew the old question of the meaning of Being, the most universal reality that is presumed in every "is." Heidegger was alienated by the dehumanizing effects of modern technology and thought past it through Being. He highlighted topics such as Being-toward-death, authenticity, and care. For him, care is the central quality of Being. Being was experienced by those astronauts whose trips into space did not make them more domineering but awakened them to a reverent sense of the universe's wonder. Back home, Being can be experienced in the background of earth's lifeworld, and it flowers in moments of meditative or ecstatic wonder, joy, and caring, when, as they say in Zen, "you see your original face."

Heidegger's profound insight into the nature of the realm of Being prior to the concepts of the object and the subject in western metaphysics has stimulated many subsequent thinkers. For him, the industrial object exits only as it stands against the subject. In his essay "The Age of the World Picture," the object for industrial consciousness is no longer that which is presented by Being, by existence itself. The object has become that which "stands against" (the English word *object* is translated from the German *der Gegenstand,* or stand-against). Objects are pictured as what stand against humans, what are secured as real "things" that are binding for all. But the process of deciding what counts as an object is a process of "objectification," stripping the world and soul of the presence of Being (Heidegger, *Question,* 150).

The main premise of this process of objectification that makes the modern world is that objects are defined or represented by subjects, beginning with Descartes. Subjective minds determine what counts, what criteria meet the definition of objects, and relegate the rest to trivial subjective feelings or foolish illusions. In other words, the purpose of objectification is human mastery of the world by definition and representation.

Representing sets up a modern domain of reality, but it is a prior projection of a fixed ground plan in order to secure a realm of objects (118). Much of this ground plan is called "physics," and we do not notice that adherence to it contains humanity in subjectivity. In modernity, the very essence of humanity theoretically changes into the metaphysical construct of the subject.

What we think of as objects are definitions of thingness that are collectively determined by subjects. Many post-Cartesians dispensed with God and relegated the possibility of certain knowledge to the supreme subject's rationality. Soul, in this newly constructed world, is restricted to subjectivity and denied to the external world of objects, which thus becomes available for mechanistic dominion.

Heidegger invited readers into *Dasein,* or "Being-in-the-world": "In saying 'I,' Dasein expresses itself as Being-in-the-world" (*Being* 368). Instead of the industrial ego as subject standing against a world of neutral, dead objects, Being-in-the-world is the more primordial ontological situation of always existing by means of participation in the environment that makes existence possible, from the origin of the universe, gravity, planets, elements, atmosphere, life, and our species' consciousness. What are commonly known as "things" are entities present-at-hand in the world, but this need not be in a narrowed mode of objectivism. Humans are not isolated subjects, but *Dasein* ("being-there"), a personal focus of Being becoming present in this world.

We live in the world of Being always and already, as whenever we use a form of the verb "to be." But to understand Being, one must reflect on and perceive his or her own Being (*Being* 27). Being is not an entity any more than God is a thing, subject to thinglike rationalistic analysis. *Dasein* is each person's Being-in-the-world not as a subject but as "Being-ontological," or open to the field of the deepest conditions of existence itself, prior to cognitive, metaphysical assumptions of thinking. The *ontological question* leaves aside the old ontological argument as too rationalistic and asks, as a meditative reflection, What makes existence itself possible? or, Why is there something rather than nothing?

Being-in-the-world opens to primordial experience of existence that appears. Rather than seeing yourself as an interior subject who views the sun setting on a world of external objects, watch a sunset and let yourself feel the earth rotating away from the sun, sending visible energy and making shadows move along the globe. Feeling yourself to be part of this solar system's light energy and earth rotation is an ageless experience of being in existence. You are alive only because of this system of existence—light energy from the sun sending just the right warmth to support life on this planet. The major characteristic of Being is care (*Being* 155), disclosed in the call of conscience (314), and Being-in-the-world includes Being-with-others in care. *Dasein* as a person's center of a lifeworld is far more inclusive than the concept of subjectness.

In the late 1930s, Heidegger wrote more specifically about technology, stressing that the essence of technology is not at all technological because the essence of industrial technology is in its Enframing (*Question* 28). By "enframing," he means setting up frameworks of systems such as the physical sciences, calculation, information, and causality. These are ways to enable the subject to control the world while at the same time repressing, concealing, or denying Being. Technological frameworks already express and grasp the modern mind in a way that is more threatening than its machinery, because they blind us to the presence of Being:

> The threat to man does not come in the first instance from the potentially lethal machines and apparatus of technology. The actual threat has already affected man in his essence. The rule of Enframing threatens man with the possibility that it could be denied to him to enter into a more original revealing and hence to experience the call of a more primal truth. (28)

Enframing thus narrowly restricts the modern mind within its horizon, making it difficult to imagine another reality. Combined with technology nihilism became explosively deadly.

Heidegger argues that Nietzsche heightened modernism immensely by discarding Being, seen as God, and replacing it with a subjective assertion of the will to power: "[I]n the age of the consummation of Western metaphysics in Nietzsche, the intuitive self-certainty of subjectness proves to be the justification belonging to the will to power" (91). This heightened modern nihilism.

Heidegger's thought is not an apology for God but a search for a more *primordial experience* of the reality or presence of Being, the power behind the many images of divinity. This approach can lead to a postmodern spirituality that is consistent with the best in science, if it includes a deep analysis and critique of both science and religion. Seeds of this can be seen in Paul Tillich's rejection of the anthropomorphic father God of Christianity for the language of the "Ground of Being." Truth comes from unconcealment, out of a meditative reflection, not only from the logic of subjective certainty. Transparency to the presence of Being provides a path to a new ontology, a new way of Being-in-the-world that leads past modernity to a more authentic existence.

Opening technology down to its deeper ground is not at all to reject technology. Rather, it is to open up a new cultural paradigm altogether, one that stops the excessive calculation, domination of nature, commercial onslaught, and repression of what industrial culture marginalizes. These bold changes cannot occur without a deep transformation of thinking about the background of reality, the ground of existence, the place where the universe cares. This is the ontology of spirituality, not materialism.

When we cease to think of ourselves as subjects seeking to control a world of objects and release into the receptiveness, the "letting be" of Being-in-the-world, suppressed forces will be freed and souls released from the alienation, despair, and nihilism built into industrial society. Clearing space for Being to make itself present (including in appropriate technologies) will cultivate serenity. Heidegger says, "The Serene [serenitas] preserves and holds everything in tranquility and wholeness. The Serene is fundamentally healing. It is the holy" (Existence 251).

This radically transforming turn away from subjectivity to Being is a necessary element in unmasking the enchantments of technology. Since it is very abstract, however, it needs to be embodied and lived. It emerges in the therapeutic experiences of depth psychology and meditative practices. Therapy leads into the soul's depths and heals the wounds that block openness to Being's serenity. Meditation calms the busy ordinary mind and helps lead it back to the Original Mind, another word for ineffable Being, full of care, compassion, and a more robust, poetic truth. Thus humans can "dwell poetically on Earth."

In "The Question of Technology," Heidegger challenges the instrumental view that technology is a means to an end and even the anthropological view that it is a human activity. More deeply seen, technology is a mode of Being revealing itself. The role of humanity in the unconcealment of that which is is to participate with the gathering and enframing that emerges as technology. The machine is not an isolated thing but rather part of a much larger way of making present the real in creations. Nor is technology only the result of the process of cause and effect. The ontological question instead asks, Whence comes the realm of cause and effect?

To reject the domination of technology and be open to the question of Being puts technology in a completely new perspective, since it is not seen now as simply the product of human will over matter, but as the disclosure of Being. Technology itself is finally not the greatest threat to humanity. No matter how deadly the machine, chemical, or genetic code, the lethal threat lies in the artificially constructed role of the human ego, the failure to heed the call of a more primal truth. Escaping that, a new consciousness arises. Plato called it *ekphanestaton:* "that which shines forth most purely" (Heidegger, *Question,* 34). If humanity truly opened itself to that shining primal truth, we would not let technology become a threat. There would be a deeper release from our metaphysical foundations of technology that would result in new ethical restraints because we have allowed Being to disclose itself more fully.

The technological quest for power is so boundless that it is enchanted not by a rational purpose—a sensible acquisition of adequate power—but by the insatiable grasping for power by a shriveled and nihilistic ego. Nietzsche's will to power has gone beyond itself to become the will to will, the drive to gain power for the sake of power, not for any higher purpose. Although religious ethics should clearly challenge this infinite craving, other religious themes (such as dominion) have failed to prevent this mad technological excess of expansive power that has allowed so many efforts at becoming godlike. This metaphysical background of the will to power, not technology itself, is the essence of technology. Technology is thus a one-dimensional, tool-oriented quest for power that is blind to its side-effects, its nihilism, and its immersion in Being-in-the-world. The whole world would be seen as our way of participating in the great mystery of existence, rationally, gratefully, and compassionately enchanted by the wonders of Being-in-the-world.

Deep Technology

What we need is a wide-ranging, healing "deep technology." Technonihilism would be recognized and countered with preventive measures, such as improving the quality of life, as opposed to frantic materialism. Care, rather than competition, would be elevated. The meanings of life would be consciously examined and critiqued if crudely technological and aggressive. We would recognize that technology is not self-regulating, neutral, or objective but comes packaged with values and passions that need critical analysis. Scientists and technologists would recognize that they already have nonobjective guiding principles and consciously refine these to the highest level. Personal life would not be shoved aside by industrialism's educational and work pressures but would be seen as a realm for increased self-awareness and cultivation of refinement. Awareness of one's own unconscious would be taught and encouraged. Raw desires emerging from the depths would be refined so that egotism and domination of nature and others would be rejected in favor of participation in soul-in-the-world and Being-in-the-world. Thus regulation of technology would flow naturally from more refined souls. Technology's wonders would not be taken for granted and greedily abused but would be seen as ways that soul-in-the-world gives powers that need guidance. Technological utopianism would be tamed by clarity about the shadow dangers.

Industrial society's development of technology has been at one extreme a kind of free-for-all of democratic societies welcoming the clever inventions of anyone, so we get inspiring garage startups that end up as major computer corporations. This dream ballooned in Silicon Valley computer projects until the bubble of that development burst, and that progress slowed down. At the other extreme we have tightly controlled government research programs, largely military. Whether the goals are profit or conquest, too much technological focus is destructive.

It is time for scientists, engineers, and all technically skilled people to begin seeing themselves as a new category of professionals, similar to doctors, lawyers, and clergy. Rather than pursuing technosuccess at any cost, in a race to be first, to get rich, or to outsmart the enemy, technologists, from car mechanics to space scientists, need to enlarge the vision of their work to include professional ethical standards. As Bill Joy says, scientists and engineers should adopt a strong code of ethical conduct, resembling the Hippocratic Oath (12). International organizations could take the lead in promoting ethical scientific behavior. Physicians for Social Responsibility, for example, already takes strong positions on nuclear disarmament and gun control.

Professionalizing technological development would deepen the work by acknowledging the purposes, motives, social context, and urgency of staving off disasters. Rather than retaining the model of "objective knowledge," value neutrality, and unlimited research that predominates, technological development should raise itself up to a professional level by expecting colleagues to acknowledge and take responsibility for the unconscious dynamics at work under the surface. Rather than allowing a crude "anything-goes" or even nihilistic style of developing technology, we should expect scientists and technicians to take refined ethical responsibility. Rather than irresponsibly cultivating hyperegos who play god for glory in a race of progress and personal gain, let us expect technoculture to police itself, limit dangerous technologies, take ruinous consequences seriously, and prevent disasters. If they cannot, then international organizations should start demanding such professional commitments.

Deep technology would acknowledge its place in soul-in-the-world and Being-in the-world. Car designers know they are building sexy aggression machines, so why should they and their marketing executives not be held professionally responsible for the soul-in-the-world that they put on wheels? Deep technology would expect space engineers to ethically justify their expensive, dangerous technologies with ethical rationales, not simply a quest for godlike knowledge and power. Deep technology would hold nuclear engineers and policy makers responsible for the cancer and birth defects in the wake of radiation spread around by depleted uranium in weapons.

Bill Joy is an example of a deep technologist. He knows that technoculture is driven "by our habits, our desires, our economic system, and our competitive need to know" (11). He knows that we have opened several Pandora's boxes of species-threatening technologies, and that to save our species, we are being forced into policies of "relinquishment: to limit development of the technologies that are too dangerous" (11). How intelligent is it to let our unconscious or denied crude desires drive us to extinction? Are we not smart enough to control our own self-destruction, because like helpless addicts, we cannot restrain our enchantments? Deep technology would require a fundamental change in technological thinking, a psychological and spiritual refinement. It would require more self-awareness, more self-restraint, a refined sensitivity to enchantments, and more participation in Being-in-the-world.

The Enchanted Ego

As we have uncovered the many enchantments of technological culture and the deeper participation in Being-in-the-world that underlies the superficial subject/object dichotomy, we have regularly heard people proclaiming the godlike quality of the industrial ego. When a robot maker says, "'It's tough being a god'" (qtd. in Lyall), we have to wonder what audacious consciousness this expresses. As modernity developed, godlike powers were slowly transferred from the heavens to the mortal egos of industrial humans. Descartes's God was a guarantor of truth for his system, but as the industrial metaphysic gained strength, it came into increasing conflict with traditional church dogmas. Although private religious belief remains strong in nations such as the United States, the public influence of western religion has reached a turning point. From cell phones to space missions to Mars, the dramatic impact of modern technology has had an immense effect on western consciousness. Outside fundamentalist circles that struggle to retain a shred of transcendence in public life, the traditional sense of God as active in history has faded into the background. The existence of the traditional qualities of the omnipotent, omniscient, compassionate Creator, the ultimate reality in traditional religions, is no longer taken for granted.

But godlike qualities have not disappeared. Humans have a hunger for transcendence, and while they may suppress and deny it, it reappears in new ways. Those enchanted by the powers of technology are turning over godlike qualities to the human subjective ego. Samuel Butler saw in the nineteenth century that machines were turned into ersatz love objects, fetishes, and almost gods (Mumford, *Pentagon*, 196). Freud and Marx denounced religion as illusion. But technological culture is practicing an enchanted cryptoreligion just under its surface. In 1900, Henry Adams grieved that Americans felt no goddess as powerful as the steam engine. H. G. Wells named a book *Men Like Gods*. Bucky Fuller wrote about *No More Secondhand God*. Isaac Asimov entitled a robot story *The Gods Themselves*. "'You have to play God, and that's not easy,'" said one Biosphere scientist (qtd. in Appenzeller 1370). Speed elevates its fanatics into a frenzy of feeling transcendent and sublimely godlike. Kneeling at the altar of speed, hungry for unlimited power, drivers see the extravagant culture of speed as a God-given right and assurance of technological superiority. James Watson, the codiscoverer of the DNA double helix structure, said, "'If we don't play God, who will?'" (qtd. in Moore, "Watson"). The omnipotent powers of the old God in the heavens are being claimed by technoculture.

Lindbergh saw this. When he first watched a parachute jump, he saw the brave jumper as "a man who by stitches, cloth, and cord, had made himself a god of the sky for those immortal moments" (255). "What freedom lies in flying!" Lindbergh exclaims, "What godlike power it gives to man!" (94). Flying is a huge leap beyond utilitarian practicality to the edge of the infinite. Speeding along above the earth, flight easily inflates the ego. Wild fantasies of space stations or remote planetary utopias (which ignore realities such as deadly cosmic radiation) could only be taken seriously by inflated techno-egos. Fanatics such as Marinetti inflate their egos to assume godlike powers and rights, brushing aside ethics in the name of technological domination. Freedom to dominate is taken for granted. The question why the largest moving machine in the world, the *Titanic*, was named after a tribe of violent Greek gods, the Titans, was tamely dismissed. The name still echoes with the blinding pride of its technological grasping for utopian power, as well as the tragedy of its failure. Enchantments grab us, and we play out their dramas—control, sex, power—through our psyches and technologies. Technodreamers absorb godlike powers—but not divine wisdom—into their own egos.

Like Prometheus stealing fire from the gods (now enshrined in Rockefeller Center), Faust selling his eternal soul for illusory pleasures, and Dr. Frankenstein recklessly creating a life that becomes a beastly horror, industrial society's dreamers have not thrown gods off the stage but have identified with omnipotent divine powers. They have sought to become godlike themselves, recklessly speeding, polluting, cloning, and exploring the heavens with flagrant disregard for the pretentious qualities of their unconscious motives. Icarus symbolizes the age-old yearning for *apotheosis*—becoming godlike, omnipotent, omniscient, surveying the world from above, the ultimate creator and savior. Like Icarus, the modern subjective ego is reaching for the stars and risks falling, for it has lost its soul.

Android robots are imaginative players in the Pinocchio Project, seeking to find absolute reality in machinery. Some robot makers fantasize themselves creating a new species. They make (usually cinematic) gold-plated images of the inflated techno-ego, idols of the Absolute Machine, Robogod. It does not function very well, but it seeks to demonstrate the totality of the machine metaphysic. Futuristic predictions that robots will exceed humans in fifty years are enchanted visions of the bloated ego of industrial society. Reductions of the human mind to fit the limits of computer language so mind can be "downloaded" into machines is a terribly shallow estimate, cutting the mind off from its deeper soul, where such enchantments originate.

This scorn for the natural mind is part of the dream of computers and robots becoming godlike, as in science-fiction fantasies. Science fiction is a cryptoreligion that offers endless dreams of a total mechanical metaphysic and human omnipotence. But it is inhabited by a few heroes surrounded by stunted cyborgs in fairy-tale melodramas. They survive only by dreamy technomiracles. Worse, science fiction offers no answer to its moral ambivalence. Will the dark side of human nature and its technology be defeated by the light side or not? There is no assurance, so this cryptoreligion is full of robotic Terminator anxiety and Darth Vader fear. Its only promise is the dream of progress, a future of utopian technologies and wildly inflated human egos unaffected by the realities of space-time travel. The dethroned God has been drawn down into the pumped-up techno-ego, nicely symbolized by metallic android robots with glowing red demonic "eyes."

The industrial ego has become a *hyperego.* Its strong will to power has used reason to provide the upper crust of the world's population with comforts that would dazzle ancient kings. And it has come to take them for granted, indulging in road rage and deficit spending if anything gets in its way. "Freedom" has become a code word for the hyperego's right to use technology to gain excessive power, narcissistic wealth, horrendous weapons, and self-indulgent pleasures. Ethical responsibility is too easily shunned by career-hungry, materialistic hyperegos willing to abuse corporations and governments for greedy self-aggrandizement. The cult of celebrity, inflated by media technology, treats performers and athletes like semidivine idols, and hordes of youths with no better models lust to imitate their shallowness. Conversely, the faith in technology that the hyperego demands is a shriveled yet dangerously extravagant faith. In *The Will to Technology and the Culture of Nihilism,* Arthur Kroker says, "A *hyper-religion,* the will to technology requires an act of faith in the efficacy of technology itself as the ritual of admission into its axiomatic procedures" (29). The subjective technological ego seeks to absorb godlike omnipotence over its kingdom without godlike wisdom, and thus indulges in crude enchantments. Bill Joy says: "The truth that science seeks can certainly be considered a dangerous substitute for God if it is likely to lead to our extinction" (11). It is curious that the industrial ego does not restrain itself within the limits of its rational logic but repeatedly makes utopian proclamations, dreamy visions, and claims to godlike absolute power. Why would this supposedly rational mental tool so easily become passionate and imaginative? Not only do such barely suppressed enchantments easily surface, but they bring up the particularly tempting desire to be all-powerful, like the traditional western God.

Western theologians such as Anselm defined God as the greatest conceivable being, and Aquinas added themes such as the first cause (or prime mover) and intelligent source of universal order. It is one thing to discard such a God in order to rebuild a new world order based on science instead of divine authority. It is another for the egos of the new scientists to begin to absorb the rejected qualities of the divine creator—omnipotence and omniscience—into themselves.

There is a lot of casual talk about the Internet as God, robotics engineers seeking immortality for themselves, and certain video games as "god games," where players can gain maximum control, create moral codes, and look down from the heavens on an imaginary world they created, acting like a "virtual god." One veteran game designer said, "It's kind of fun to be worshipped" (Totilo). What an enchanted leap from reason to divinity!

This may be a compensation for discarding God from industrial metaphysics, showing the need for some relation to the divine, no matter how crude. Or it may be the simple lust for ultimate power gone wild in the hyperego. But playing God is a dangerous, inflated enchantment that too often seems to permit beneficiaries to excuse themselves from mortal humility and moral limits. Most important, to naïvely grab the omnipotence and omniscience of God and totally neglect the overarching theology of the divine compassion, justice, humility, forgiveness, and peace is a blind, self-serving enchantment.

But there is a glimmer of hope. Instead of identifying with gods, many of the real astronauts, as we saw, lost their hyperegos when they saw the reality of the earth from orbit. The U.S. astronaut James Irwin spoke of seeing the earth on the way to the moon: "Seeing this has to change a man, has to make a man appreciate the creation of God and the love of God." He saw the moon not as an object of conquest but as a "very holy place" (Kelley 38). Rusty Schweickart said that in space he was "touched by God" (Kelley 144). A major shift in the technological ego emerged on those space missions, pulled away from the techno-ego's inflated pride in mastery toward a humble sense of our cosmic place. It awakened an obligation to protect the earth's environment rather than indulge in dreams of space conquest.

Edgar Mitchell's radical epiphany transformed him from a utilitarian astronaut to a highly spiritual mystic. What he experienced in space shook his conventional industrial consciousness to the core: "[A]n infinite sea of unstructured energy potential from which the universe arose" (157). He saw that creation was not a random, meaningless event but rather that it intentionally arose from this underlying "quantum potential" (157). The universe is more than whirling matter and glowing energy, for mind is an integral part of this system. Mitchell refuses to name the divine in any anthropomorphic traditional sense of God as the old man in the clouds. For him, "the origin of all religions is rooted in the mystical experience" (137). He sees that our local gods are too small, for they really fill the entire universe. And our science is too narrow, for it excludes the awesome cosmic power that brought the universe into being and the "eternal, connected, and aware Self experienced by all intelligent beings" (216).

This awareness is always available to humans, and surprisingly so in space travel. On the frontier of space, cosmic experiences are bursting out of the capsule of technological metaphysics. What Mitchell and others experienced is called by many names: the presence of Being, the holy source of all existence, or the sound of one hand clapping. Astronauts like Mitchell are opening out into a new spirituality where the ego does not absorb divine powers but serves the highest power that fills the universe and our souls with wonder and peace. Serving this vast, gentle wind is an enchantment with Being, dancing in your own star-energy.

New Gates

What lies below the subject/object dualism? You can unlock many new gates to this timeless secret garden by peeking into our repressed perceptions of soul-in-the-world and Being-in-the-world, beginning with questions to be meditatively pondered at a series of gates. At the first new gate, you would become aware of our enchantments. Rather than fencing our industrial consciousness off into a narrow rationalistic, utilitarian, and consumerist mindset, you can open up to the depth psychology that explores deeper motives. When are we acting out of compassion, and when are we acting out of the will to power? How is our reasoning interacting with our passions? How does that affect our technologies? We can never rid ourselves of all enchantments, and we would not want to, but we can learn to differentiate between positive and negative enchantments and take ethical responsibility for them.

At the second gate, you actively cultivate your dormant sense of Being-in-the-world. In a quiet time, ask yourself: What are creations? Is it possible to separate a creation from its context, stand detached against it, and use it for human purposes only? What is the life-support system that makes that item possible? What raw materials were used to make it? Where did they come from? Do you recall that each atom in your body came through a star exploding ten billion years ago? What is the source of the entire lifeworld's systems? What animals or plants did you eat today? Your computer's plastic comes from oil, derived from ancient decayed and buried organic matter; your home's wood comes from the system of seeds and trees bathed in light, water, and soil; and so forth.

At the third gate, ask yourself: What makes possible soul-in-the-world's consciousness? Is it confined to your "innerness" alone, or is it part of a vast, collective consciousness, extending beyond our time and space? What makes the brain and its life-support system possible? Are human feelings simply the by-products of the brain, or do they have their own being, partially related to the brain? What makes anomalies such as near-death experience possible? Do other life forms also have soul-in-the-world? How do you relate to pets and wild animals? Are these soul events random, meaningless phenomena evolved by chance in our enormous, primeval universe? Or are they part of a cosmic force, as religions believe, that is creative, compassionate, and healing?

How do you feel when you see pictures of Earth from outer space? Do you recall that, at light speed (which would destroy a human body), it would take a hundred thousand earth years to cross our galaxy? Does this kind of power make you feel glad to be on this safe planet? Why do we treasure love and justice so much? Why are we not simply a planet full of brutal beasts doing nothing but attacking and eating each other? How do your answers to these questions affect your sense of your Being-in-the-world? Do you feel a calming deep mystery, a bottomless well of existence that hosts your brief time on its surface? If so, what kind of person-in-the-world do you want to be in view of that? What kinds of technologies do you want to support?

At the fourth gate, ask yourself: What kind of Being-in-the-world is it to exit the race for maximum speed? Quick. Hop in your car and escape to a retreat center for a few days. Ask your hosts how often most people spend the first day or so sleeping, just to slow down from the normal hectic pace. Perhaps in a retreat cabin facing a pond, open up to the blissfully quiet and still pond's source bubbling up deep below the surface. Sit warmed by the campfire's burning wood of trees that are older than our species, growing on hills older than the tree's species. Then slip into the soul-in-the-world reverie of peaceful thankfulness for your niche in this world. Then, as the daylight fades, look up as the moon cycles overhead for the billionth time. What's the rush? What kind of technologies does such a retreat experience suggest?

At the fifth gate, ask yourself: What kind of Being-in-the-world is aware of its utopian dreams? If I see myself as a character in the dramas of utopian adventures, or a viewer enchanted with its excitement, it is part of my world, not an external book or movie. From paradise islands to science-fiction adventures, utopian dramas are symbolic of deep currents in a culture's soul-in-the-world.

Reading about or watching the *Titanic* again, you may feel drawn into the confident heroic reach for technological dominance of the ocean, proudly topped off with luxurious sweets for the upper crust. You may be pained by the blind reckless illusion, casting fifteen hundred souls to their freezing, sinking deaths. Or you may feel the chill of death, sharing the soul-in-the-world of the drowning victims of this nautical *hubris*. In today's utopia, large luxury yachts lumber around, and speedboats and personal jet-skis zoom around our oceans, lakes, and rivers, gasoline fumes smoking, and noisily annoying slower boaters. By contrast, casually swim a pond, sail a lake, or canoe a river. Hear the surging soul of the water while dancing on the waves, the billowing winds, the splashing, life-giving waters, billions of times older and larger than any of our machines, the waters springing up from the timeless, bottomless well of life itself.

At the sixth gate, what kind of Being-in-the-world emerges in the space above the earth? While flying, can you look out the airplane's window without being absolutely astonished at both the powerful technologies of flight and the tiny green farms below, the high snowy mountain peaks, the vast ocean, the vast drifting clouds ahead, and the sunbeams shooting over the curved horizon of the turning earth for the trillionth time? While human space travel captivated our enchantments for a while, its cost, danger, and wastefulness soon challenged hopeful dreams. But the few returning astronauts have become leaders of our awakening enchantment with the reality of the earth's awesome majesty and cosmic wonder, transforming them from utilitarian engineers to mystics. They bring back to earth an exciting and urgently needed new spiritual and ecological sense of Being-in-the-cosmos. Soak up the implications of some outer-space photographs and be lifted far above our old vision of the world as a dead object.

At the seventh gate, you become a robot. What kind of Being-in-the-world is this? Suspending disbelief, you imagine yourself to be a high-tech android puppet participating in the world of pure mechanism, nothing but a machine guided by clever computer programs. How does that feel? Are you a master or a slave? Being a robot gives you far more mechanical power than humans. You can turn into a lethal weapon, dispensing destruction, or a functionary with a vast memory and automated systems far beyond human scale. You could be a surgeon saving lives, play chess, or join a bomb squad defusing terrorist weapons.

Next, you are purchased by a group that programs you to assassinate politicians. You efficiently help with a couple of fascist political coups. Then you begin to process what your programmer calls "feelings," and things change. Now you feel "happy," "sad," or "angry." Before long you feel "male," or "female," and want "sex," and so you make yourself a partner. Then, after reading the *Pinocchio* story, you desire nothing more than to become a real human. But you cannot transfer their programs, since they are not compatible. So you decide to maximize all your powers and become what humans call a "god" that they will have to obey. That will show them who is boss.

Back down to earth, you can begin to reflect on what enchantments, such as the above, inhabit the worlds of robots, both functional and android. What happens when humans begin to develop the mechanical aspects of Being-in-the-world without being conscious of what they are doing. How are we repressing the imaginative, fantasy element, the desires, the dreams, the soul-in-the-world purposes prior to the logical calculations that make machines? How much cost, danger, and control should humans grant robots? The mechanical/electronic dimension of the universe is certainly powerful and fascinating. As the most evident to the senses, it is easily manipulated. But what mysteries does it hide? What other factors also comprise Being-in-the-world, and how do they relate to the mechanical/electronic? Is "energy" purely mechanical and electronic? Or does it have another, more subtle aspect, that manifests in various kinds of healing, for example?

The key question at the last gate is: Did all of existence, the entire universe, cause and define itself? Or does existence have a prior foundation, a source greater than all its creation, which is both transcendent to it and present in a mysterious primordial reality evident in the material, the energetic, and the soulful phenomena of our glimpses of outer space? When technology based on science emerged in history, religious institutions invested in an authoritarian positive answer to the second question resisted the new technology's explorations, because it challenged their worldview.

So new scientists and inventors had to reject its old theology in order to honestly formulate the new scientific and technological metaphysics. This latter metaphysics still resists the religious worldview, which, at a literal level, remains archaic, though still meaningful to many. But as we have seen, the technological metaphysics of the subject and object is not a pristine structure. It is thoroughly enmeshed with the purposes, desires, and enchantments it seeks to expel from its "objective" worldview. When this happens, the validity of the traditional metaphysics of industrial society comes under question.

This view by no means calls for a return to old religious metaphysics. But it does at least suggest that there is more to existence than the subject/object metaphysics has declared acceptable. And if its nihilism and other failures of technology are to be solved, there is an urgent need to open to the intuitive perceptions that some call Being-in-the-world. Other names may be just as suitable, but the point is the experience and its meaning. This experience is actively, energetically alive in the enchantments of technology, such as the urge of some of its creators to see themselves as gods. This can offer a fresh opening to a new ontology, a new foundation for a transformed culture that blends the best of science and technology's achievements with the numinous presence, the bottomless well, the eternal ground behind ethics, the background of empirical reality, the enchanted place where the sky breathes.

REFERENCES

Abel, Lynette. "How Contempt Fuels Rage on Road." *Ithaca Journal,* August 30, 1997, 8A.

Adams, Henry. "The Virgin and the Dynamo." In *American Poetry and Prose.* Vol. 2. Ed. Norman Foerster. 1221–26. Boston: Houghton Mifflin, 1962.

A.I. (Artificial Intelligence). Dir. Steven Spielberg. Dreamworks, 2002.

Anderson, Susan H. "The Pioneer of Streamlining." *New York Times Magazine,* November 4, 1979, 92–101.

Appenzeller, Tim. "Biosphere 2 Makes a New Bid for Scientific Credibility." *Science,* March 11, 1994, 1368–70.

Aristotle. "Metaphysics." In *Introduction to Aristotle.* Ed. Richard McKeon. 238–96. New York: Modern Library, 1947.

Arnheim, Rudolf. *Art and Visual Perception.* Berkeley: University of California Press, 1974.

Aronowitz, Stanley. *Science as Power.* Minneapolis: University of Minnesota Press, 1988.

Asimov, Isaac. *The Gods Themselves.* New York: Ballantine, 1972.

———. *I, Robot.* New York: Signet, 1950.

St. Augustine of Hippo. *Confessions.* Trans. R. S. Pine-Coffin. New York: Penguin, 1979.

Avens, Robert. "Reflections on Wolfgang Giegerich's 'The Burial of the Soul in Technological Civilization.'" November 2002. <www.cgjungpage.org/pschtech/giegerich1.html>.

Bachelard, Gaston. *L'Air et les songes* (Air and dreams). Paris: Corti, 1943.

———. *The Poetics of Reverie.* Trans. Daniel Russell. Boston: Beacon Press, 1969.

Bacon, Francis. *The New Atlantis.* October 2003. <www.luminarium.org/sevenlit/bacon>.

———. "*Novum Organum* I: Aphorisms." In *Selected Writings of Francis Bacon.* 461–52. New York: Modern Library, 1955.

Bacon, Roger. *Perspectiva.* Trans. David Lindberg. New York: Oxford University Press, 1996.

Bailey, Lee W. "Giambattista Della Porta." *Magic Lantern Bulletin* 18.3 (November 1988): 2–3.

———. "Religious Projection: A New European Tour." *Religious Studies Review* 14.3 (July 1988): 207–11.

———. "Robogod: The Divine Machine." *Artifex* 7.3 (Fall 1988): 15–24.

———. "Skull's Darkroom: The *Camera Obscura* and Subjectivity." *Philosophy of Technology* 6 (1989): 63–79.

————. "Skull's Lantern: Projection and the Magic Lantern." *Spring: A Journal of Archetype and Culture* (1986): 72–87.

Bailey, Lee, and Jenny Yates. *The Near-Death Experience: A Reader.* New York: Routledge, 1996.

Bair, Dierdre. *Jung: A Biography.* Boston: Little, Brown and Co., 2003.

Baker, Peter. "First Space Tourist Enjoys $20 Million View." *Ithaca Journal,* May 1, 2002, A1.

Balsley, Gene. "Detroit Versus the Hot-Rod." In *Visions of Technology.* Ed. Richard Rhodes. 193–94. New York: Simon and Schuster, 1999.

Barfield, Owen. *Poetic Diction.* Middletown, Conn.: Wesleyan University Press, 1928.

————. *The Rediscovery of Meaning, and Other Essays.* Middletown, Conn.: Wesleyan University Press, 1977.

Barnes, John. *Optical Projection.* St. Ives, U.K.: Barnes Museum of Cinematography, 1970.

Barnes, Julian. "Creating the Soul of a Robotic Dog." *New York Times,* February 4, 2001. <www.nytimes.com/2001/02/04/technology/04TOYS>.

Barnouw, Erik. *The Magician and the Cinema.* New York: Oxford University Press, 1981.

Barrett, Greg. "Hummer vs. Hybrid: Cerebral Tug of War." *Ithaca Journal,* September 17, 2003, 8A.

Barrett William. *Irrational Man.* Garden City, N.Y.: Doubleday Anchor, 1958.

Baudrillard, Jean. *The System of Objects.* Trans. James Benedict. London: Verso, 1968.

Bellah, Robert, Richard Madsen, William M. Sullivan, Ann Swidler, and Stephen Tipton. *Habits of the Heart: Individualism and Commitment in American Life.* New York: Harper and Row, 1985.

Bellamy, Edward. *Looking Backward: 2000–1887.* New York: Modern Library, 1887.

Benesch, Klaus. "History on Wheels: A Hegelian Reading of 'Speed' in Contemporary Literature and Culture." In *The Holodeck in the Garden: Science and Technology in Contemporary Fiction.* Ed. Peter Freese and Charles Harris. 212–24. Nornal, Ill.: Daily Archive Press, 2004.

Berg, A. Scott. *Lindbergh.* New York: Berkeley Books, 1998.

Berg, Anne-Jorunn. "A Gendered Socio-Technological Construction: The Smart House." In *The Social Shaping of Technology.* 2d ed. Ed. Donald MacKenzie and Judy Wajcman. 301–13. Philadelphia: Open University Press, 1999.

Berman, Morris. *Coming to Our Senses: Body and Spirit in the Hidden History of the West.* New York: Simon and Schuster, 1989.

————. *The Reenchantment of the World.* Ithaca, N.Y.: Cornell University Press, 1981.

Berry, Thomas. *The Dream of the Earth.* San Francisco: Sierra Club Books, 1988.

Bicentennial Man. Dir. Chris Columbus. Columbia Films, 1997.

Bijker, Weibe, Thomas P. Hughes, and Trevor Pinch, eds. *The Social Construction of Technological Systems.* Cambridge: Massachusetts Institute of Technology Press, 1994.

"Biosphere Blues." *Time,* April 18, 1994, 26.

La Biosphère d'Environnement Canada, Centre d'observation environnementale (The Canadian Biosphere of the Environment, Center for Environmental Observation). Brochure. Montréal, 2001.

Black Elk, told to John Neihardt. *Black Elk Speaks.* Lincoln: University of Nebraska Press, 1932.

"Body by Arnold." *Newsweek,* June 28, 1999, 74.

Bolen, Jean Shinola. *Goddesses in Every Woman.* New York: HarperCollins, 1985.

Bradsher, Keith. "GM Has High Hopes for Vehicle Truly Meant for Road Warriors." *New York Times,* August 6, 2000, 1.

————. *High and Mighty SUVs: The World's Most Dangerous Vehicle and How They Got That Way.* New York: Public Affairs, 2002.

Bragg, Rick. "Buffing the Olds Image at the End of Its Run." *New York Times,* January 28, 2001. <nytimes.com/2001/01/28/national/28OLDS.html>.

Breazeal, Cynthia. *Designing Social Robots.* Cambridge: Massachusetts Institute of Technology Press, 2002.

Breznican, Anthony. "'A.I.' Merges Visions of Two Great Filmmakers." *Ithaca Journal,* June 28, 2001, 7.

Brooke, Roger. *Jung and Phenomenology.* New York: Routledge, 1991.

Brooks, Byron. *Earth Revisited.* Boston: Arena Publishing, 1893.

Brown, Frederick. "Answer." In *Starshine.* 16. New York: Bantam, 1954.

Brown, Warren. "Unleasing the Beast." *Washington Post,* July 1, 2001, 1.

Brun, Jean. *Les Masques du Désir* (The Masks of desire). Paris: Buchet, 1981.

————. *Le Rêve et La Machine* (The Dream and the machine). Paris: La Table Ronde, 1992.

Burr, Roger. "Religion and Ethics Forum." *Ithaca Journal,* August 30, 2003, 6B.

Cabbage, Michael. "Building Space Station Will Be a Grueling Task." *Philadelphia Inquirer,* October 8, 2000, A12.

Campbell, Joseph. *The Hero with A Thousand Faces.* Princeton, N.J.: Princeton University Press, 1949.

————. *The Masks of God.* 4 vols. New York: Viking, 1969.

Capek, Karel. *R.U.R.* (1921). In *Of Men and Machines.* Ed. Arthur D. Lewis. 3–58. New York: E. P. Dutton, 1963.

Carlyle, Thomas. "Signs of the Times." In *The Works of Thomas Carlyle.* Vol. 27. 56–82. New York: Scribner's, 1904.

Chachere, Vickie. "Police Use of Face-Scanning Technology Called Too Intrusive." *Ithaca Journal,* July 14, 2001, 7A.

Chandra X-Ray Observatory. Harvard-Smithsonian Center for Astrophysics Homepage. <http://Chandra.Harvard.edu/x-ray_astro/darkl_matter3.html>.

Chang, Kenneth. "Can Robots Rule the World? Not Yet." *New York Times,* September 12, 2000. <www.nytimes.com/2000/09/12/science/12ROBO.html>.

Chartrand, Sabra. "A Split in Thinking among Keepers of Artificial Intelligence." *New York Times,* July 18, 1993, WK 6.

Chernus, Ira. *Dr. Strangegod: On the Symbolic Meaning of Nuclear Weapons.* Columbia: University of South Carolina Press, 1986.

Clark, Jim. "Squandering Our Technological Future." *New York Times,* August 31, 2001. <www.nytimes.com/200108/31/opinion/31CLAR.html>.

Clarke, Arthur. *Childhood's End.* New York: Harcourt Brace and World, 1953.

Cohen, Joel E., and David Tilman. "Biosphere and Biodiversity: The Lessons So Far." *Science News,* November 15, 1996, 1150–51.

Cohen, John. *Human Robots in Myth and Science.* London: Allen and Unwin, 1996.

Colin, P., and O. Mongin. *Un monde désenchanté?* (A disenchanted world?) Paris: Thèses, 1988.

Collins, Harry, and Trevor Pinch. *The Golem at Large: What You Should Know about Technology.* Cambridge: Cambridge University Press, 1998.

Collodi, Carlo [Carlo Lorenzini]. *Pinocchio.* New York: Penguin, 1996.

Corn, Joseph. *Imagining Tomorrow.* Cambridge: Massachusetts Institute of Technology Press, 1986.

————. *The Winged Gospel.* New York: Oxford University Press, 1983.

Crichton, Michael. *The Terminal Man.* New York: HarperCollins, 1972.

Crosby, Caresse. *The Passionate Years.* London: Alvin Redman Ltd., 1955.

Crosby, Donald. *The Specter of the Absurd.* Albany: State University of New York Press, 1998.

Crouch, Tom D. *The Bishop's Boys: A Life of Wilbur and Orville Wright.* New York: Norton, 1989.

Dallmayr, Fred. *Twilight of Subjectivity.* Amherst: University of Massachusetts Press, 1981.

Daly, Emma, with Andrew Revkin. "Oil Tanker Splits Apart off Spain, Threatening Coast." *New York Times,* November 20, 2002, A6.

Daly, Mary. *Gyn/Ecology: The Metaethics of Radical Feminism.* Boston: Beacon, 1978.

D'Aguillon, François (Aquilonius). *Philosophia iuxta ac mathematicis utiles* (Philosophy and related practical mathematics). Book 1, prop. 42. Antwerp: Plantin and Jo. Moneti, 1613. Trans. Herman Hecht in "The History of Projecting Phantoms, Ghosts, and Apparitions." *Optical Magic Lantern Journal* 3.1 (February 1984): 2–6.

Da Vinci, Leonardo. *The Literary Works of Leonardo.* 2 vols. Ed. Jean Richter. Berkeley: University of California Press, 1970.

Davis, Erik. *Techgnosis.* New York: Harmony Books, 1998.

Demski, William. "Kurzweil's Impoverished Spirituality." In *Are We Spiritual Machines?* Ed. Jay Richards. 98–115. Seattle: Discovery Institute, 2002.

Denton, Michael. "Organism and Machine: The Flawed Analogy." In *Are We Spiritual Machines?* Ed. Jay Richards. 78–97. Seattle: Discovery Institute, 2002.

Derrida, Jacques. "'Eating Well,' or the Calculation of the Subject: An Interview." In *Who Comes After the Subject?* Ed. Eduardo Cadava, Peter Connor, and Jean-Luc Nancy. 96–119. New York: Routledge, 1991.

Descartes, René. "Dioptrics." In *Descartes' Philosophical Writings.* Ed. and trans. Elizabeth Anscombe and Peter Geach. 241–56. Indianapolis: Bobbs-Merrill, 1954.

———. "Discourse on the Method" (1637) and "Meditations on First Philosophy" (1642). In *Descartes' Philosophical Writings.* Ed. and trans. Elizabeth Anscombe and Peter Geach. 5–124. Indianapolis: Bobbs-Merrill, 1954.

———. "Principles of Philosophy IV: The Earth." In *Descartes' Philosophical Writings.* Ed. and trans. Elizabeth Anscombe and Peter Geach. 229–38. Indianapolis: Bobbs-Merrill, 1954.

Descombes, Vincent. "A Propos of the 'Critique of the Subject' and of the 'Critique of This Critique.'" In *Who Comes After the Subject?* Ed. Eduardo Cadava, Peter Connor, and Jean-Luc Nancy. 120–34. New York: Routledge, 1991.

Dodds, E. R., and Morris Ginsberg. "Progress." In *Dictionary of the History of Ideas.* Vol. 3. Ed. Philip P. Weiner. 623–50. New York: Scribner's, 1973.

Donne, John. "To His Mistress Going to Bed." Online Literature Homepage. November 2004. <http://www.online-literature.com/donne/440>.

Dreifus, Claudia. "A Conversation with Dr. Marvin Minsky: Why Isn't Artificial Intelligence More Like the Real Thing?" *New York Times,* July 28, 1998, F3.

———. "A Conversation with Joseph LeDoux: Taking a Clinical Look at Human Emotions." *New York Times,* October 8, 2002, F6.

———. "Do Androids Dream? M.I.T. Is Working on It." *New York Times,* November 7, 2001. <www.nytimes.com/2000/11/07/science/ofFOER.html>.

Drengson, Alan. *The Practice of Technology.* Albany: State University of New York Press, 1995.

Dreyfus, Hubert. *What Computers Can't Do.* Cambridge: Massachusetts Institute of Technology Press, 1972.

———. *What Computers Still Can't Do.* Cambridge: Massachusetts Institute of Technology Press, 1992.

Dreyfus, Hubert, and Stuart E. Dreyfus. *Mind over Machine: The Power of Human Intuition and Expertise in the Era of the Computer.* New York: Free Press, 1986.

Dr. Strangelove or: How I Learned to Stop Worrying and Love the Bomb. Dir. Stanley Kubrick. Columbia/Sony Pictures, 1964.

Dubos, René. *The Dreams of Reason: Science and Utopias.* New York: Columbia University Press, 1961.

Dumas, Lloyd. *Lethal Arrogance: Human Fallibility and Dangerous Technologies.* New York: St. Martin's Press, 1999.

Dunn, Marcia. "Space Station Skipper Celebrates with Burger, Beer." *Ithaca Journal,* March 22, 2001.

Durer, Albrecht. "Drawing" (1527). In *Art and Illusion,* by E. H. Gombrich. 306. New York Pantheon/Bollingen, 1961.

Eakin, Emily. "The Pilot's Wife." *New York Times Book Review,* December 12, 1999, 14.

Eder, J. M. "The Invention of Projection Apparatus." In *The History of Photography.* New York: Columbia University Press, 1905.

Ellenberger, Henri F. *The Discovery of the Unconscious.* New York: Basic Books, 1981.

Eliade, Mircea. *Patterns in Comparative Religion.* New York: Sheed and Ward, 1958.

Eliot, T. S. "The Waste Land" (1922). In *The Norton Anthology of English Literature.* Ed. M. H. Abrams. 2582–98. New York: Norton, 1968.

Ellul, Jacques. *The Technological Society* (1954). Trans. John Wilkinson. New York: Alfred Knopf, 1964.

Emerson, Roger L. "Utopia." In *Dictionary of the History of Ideas.* Vol. 4. Ed. Philip P. Weiner. 458–65. New York: Scribner's, 1973.

"Farewell to Supersonic Travel." *New York Times,* April 13, 2003. <http://www.nytimes.com/2003/04/13/opinion13sun3.html>.

Fasching, Daryl. *The Thought of Jacques Ellul.* Lewiston, N.Y.: Mellen, 1981.

Faubion, James, ed. *Rethinking the Subject.* Boulder, Colo.: Westview Press, 1995.

Feenberg, Andrew. *Transforming Technology: A Critical Theory Revisited.* Oxford: Oxford University Press, 2002.

Ferkiss, Victor. *Technological Man.* New York: New American Library, 1969.

Ferré, Frederick. *Hellfire and Lightning Rods: Liberating Science, Technology, and Religion.* Maryknoll, N.Y.: Orbis Books, 1993.

Ferris, Timothy. "A Space Station? Big Deal!" *New York Times Magazine,* November 28, 1999, 124–29.

Feuerbach, Ludwig. *The Essence of Christianity* (1841). Trans. George Eliot. New York: Harper and Row, 1957.

Flink, James. *The Car Culture.* Cambridge: Massachusetts Institute of Technology Press., 1975.

Flint, Jerry. "As Useful as Rust, Tailfins Were an Outrageous Rage." *New York Times,* October 16, 1977, Auto Section, 3.

Florian, Jean-Pierre. "La Singe qui Montre la Lanterne Magique." In *Fables.* 63–66. Paris: Houdaille, 1843.

Foerst, Anne. *God in the Machine: What Robots Teach Us About Humanity and God.* New York: Dutton, 2004.

Fortmann, Han. *Als Ziende de Onzienijke* (As seeing the unseen). 4 Vols. Hilversum, Netherlands: Gooi en Sticht, 1964–74.

Frazer, James G. *The Golden Bough* (1922). Abridged ed. New York: MacMillan, 1971.

Fredericks, Casey. *The Future of Eternity: Mythologies of Science Fiction and Fantasy.* Bloomington: University of Indiana Press, 1982.

Freud, Sigmund. *The Psychopathology of Everyday Life* (1901). Vol. 6 of *The Standard Edition of the Complete Psychological Works of Sigmund Freud.* Ed. J. Strachey. Trans. J. Strachey, Anna Freud, Alix Strachey, and Alan Tyson. London: Hogarth, 1974.

———. "Some Points for a Comparative Study of Organic and Hysterical Motorparalyses" (1873). In vol. 1 of *The Standard Edition of the Complete Psychological Works of Sigmund Freud.* Ed. J. Strachey. Trans. J. Strachey, Anna Freud, Alix Strachey, and Alan Tyson. 157–72 London: Hogarth, 1974.

Fuller, R. Buckminster. *No More Secondhand God.* New York: Anchor/Doubleday, 1971.

Gartman, David. *Auto Opium: A Social History of American Automobile Design.* New York: Routledge, 1994.

Gauchet, Marcel. *The Disenchantment of the World.* Trans. Oscar Burge. Princeton, N.J.: Princeton University Press, 1997.

Geduld, Harry. "Genesis II: The Evolution of Synthetic Man." In *Robots.* Ed. Harry Geduld and R. Gottesman. 3–37. Boston: New York Graphic Society/Little Brown, 1972.

Gernsheim, Helmut, and Alison Gernsheim. "The History of the *Camera Obscura.*" In *The History of Photography.* 17–29. New York: McGraw-Hill, 1969.

Gerth, H. H. and C. Wright Mills, eds. (1946). *From Max Weber: Essays in Sociology.* New York: Oxford University Press, 1967.

Giegerich, Wolfgang. "Der sprung nach dem wurf: Über das einholen der projektion und der ursprung der psychologie" (The Leap after the throw: On the withdrawal of projection and the origin of psychology). *Gorgo* 1.1 (1979): 49–71.

Gill, Brendan. *Lindbergh Alone.* New York: Harcourt Brace Jovanovich, 1977.

Ginsberg, Morris. "Progress in the Modern Era." In *Dictionary of the History of Ideas.* Vol. 3. Ed. Philip P. Weiner. 633–50. New York: Scribner's, 1973.

Gleick, James. *Faster: The Acceleration of Just About Everything.* New York: Random House, 1999.

Glendinning, Chellis. *When Technology Wounds: The Human Consequences of Progress.* New York: Morrow, 1990.

Goldstein, Laurence. *The Flying Machine and Modern Literature.* Bloomington: Indiana University Press, 1986.

Gombrich, E. H. *Art and Illusion.* New York: Pantheon/Bollingen, 1961.

Goodman, Ellen. "America's Cultural Car Clash." *Boston Globe,* June 11, 1999, 11A.

Gorman, James. "When Nerds Collide: Bots in the Ring." *New York Times,* January 24, 2002.

Greenberg, Daniel. "Waste in Space." *Washington Post,* February 7, 2003, A27.

Griffin, David Ray, ed. *The Reenchantment of Science.* Albany: State University of New York Press, 1988.

———, ed. *Reenchantment without Supernaturalism: A Process Philosophy of Religion.* Ithaca, N.Y.: Cornell University Press, 2001.

Guy, Laurel. "Auto Biography: My Car Is My Life." Performance. September 8, 2001. Autumn Leaves Bookstore, Ithaca, N.Y.

Haraway, Donna. *Modest_Witness@Second_Millennium: Female_Man_Meets_Oncomouse.* New York: Routledge, 1997.

Harman, Willis. *Global Mind Change.* 2d ed. San Francisco: Berrett-Koehler, 1998.

Harris, John. "Camera Obscura" and "Magic-Lantern." In *Lexicon Technicum.* London: N.p., 1704.

Hawthorne, Nathaniel. "Fancy's Show Box" (1842). In *Twice-Told Tales.* 272. New York: Lupton, 1943.

Hayden, Thomas. "The Age of Robots." *U.S. News and World Report,* April 23, 2000, 45–50.

Hazen, Robert. "Battle of the Supermen." *The Guardian,* April 15, 1989.

———. "What We Don't Know." *MIT's Technology Review.* July 1997, 100.5, 23–30.

Healey, James R. "The Car's Century." *USA Today,* December 31, 1999, B1.

Heidegger, Martin. *Being and Time* (1927). Trans. John MacQuarrie and Edward Robinson. New York: Harper and Row, 1962.

———. *Existence and Being.* Trans. Douglas Scott, R. F. C. Hull, and Alan Crick. South Bend, Ind.: Gateway, 1949.

———. *The Question Concerning Technology.* Trans. William Lovitt. New York: Harper and Row, 1977.

Helton, Roy. "Machinery Has Destroyed the Peace." In *Visions of Technology.* Ed. Richard Rhodes. 142–45. New York: Simon and Schuster, 1989.

Henson, Hilary. *Robots.* New York: Warwick Press, 1988.

Hesiod. *The Works and Days; Theogony; The Shield of Heracles.* Trans. Richard Lattimore. Ann Arbor: University of Michigan Press, 1970.

Hildegard of Bingen. *Hildegard.* Ed. Robert Van de Weyer. London: Hodder and Stoughton, 1997.

"Hillary, Edmund." New Zealand Edge Homepage. July 2003. <www.nzedge.com/heroes.html>.

Hillman, James. *The Dream and the Underworld.* New York: Harper and Row, 1979.

———. *Re-visioning Psychology.* New York: Harper and Row, 1975.

Hillman, James, and Michael Ventura. *We've Had a Hundred Years of Psychotherapy and the World's Getting Worse.* San Francisco: Harper, 1992.

Hitt, Jack. "The Next Battlefield May Be in Outer Space." *New York Times Magazine,* August 5, 2001. <www.nytimes.com/2001/08/05/magazine/05SPACEWARS.html>.

Hobbes, Thomas. *Leviathan* (1651). New York: Norton, 1968.

Hoffmann, Paul. "Enchantments of a Swiss Town." *New York Times,* April 18, 1993, Travel Section, 8–10.

Holt, David. "Projection, Presence, Profession." *Spring: A Journal of Archetype and Culture* (1975): 130–44.

"Honda, Technology, Asimo." Honda Homepage. December 12, 2004. <http://www.Honda .co.jo>.

Homer. *The Iliad of Homer.* Trans. Richard Lattimore. Chicago: University of Chicago Press, 1951.

"Hummer." *Newsweek,* January 31, 2000, 8.

Husserl, Edmund. *The Crisis of European Sciences and Transcendental Phenomenology* (1954). Trans. David Carr. Evanston, Ill.: Northwestern University Press, 1970.

Huxley, Aldous. *Brave New World.* Garden City, N.Y.: Doubleday, 1932.

Huygens, Constantijn. *Oeuvres completes.* Vol. 4. Le Havre: Nijhoff, 1911.

I, Robot. Dir. Alex Proyas. Twentieth-Century Fox, 2004.

Ihde, Don. "Epistemology Engines." *Nature* 406 (July 6, 2000): 21.

———. *Philosophy of Technology: An Introduction.* New York: Paragon House, 1993.

———. *Technology and the Lifeworld.* Bloomington: University of Indiana Press, 1990.

Institute of Noetic Sciences Brochure. 2003. <http://www.noetic.org>.

Irving, Washington. "The Legend of Sleepy Hollow" (1820). In *American Poetry and Prose.* Vol. 1. Ed. Norman Foerster. 313–27. Boston: Houghton-Mifflin, 1962.

James, Leon, and Diane Nahl. *Road Rage and Aggressive Driving.* Amherst, N.Y.: Prometheus Books, 2000.

Joy, Bill. "Why the Future Doesn't Need Us." *Wired,* April 2000. <http://www.wired.com/wired/archive/8.04/joy.html>.

Jung, Carl G. "Approaching the Unconscious." In *Man and His Symbols.* Ed. Carl G. Jung. 1–94. New York: Doubleday, 1964.

———. "The Conjunction." In *Mysterium Conjunctionis.* Paras. 654–789. Vol. 14 of *The Collected Works of C. G. Jung.* Ed. William McGuire. Trans. R. F. C. Hull. Princeton, N.J.: Princeton University Press, 1953–79.

———. "Flying Saucers: A Modern Myth." Paras. 589–824. In *Civilization in Transition.* Vol. 10 of *The Collected Works of C. G. Jung.* Ed. William McGuire. Trans. R. F. C. Hull. Princeton, N.J.: Princeton University Press, 1953–79.

———. *Memories, Dreams, and Reflections.* Ed. Aniela Jaffé. Trans. Richard Winston and Clara Winston. New York: Random House, 1961.

Kasser, Tim. *The High Price of Materialism.* Cambridge: Massachusetts Institute of Technology Press, 2002.

Kay, Jane Holtz. *Asphalt Nation: How the Automobile Took Over America and How We Can Take It Back.* Berkeley: University of California Press, 1997.

Keilor, Garrison. "A Prairie Home Companion." July 7, 2001. A Prairie Home Companion Homepage. <http://prairiehome.publicradio.org/programs/2001/07/07>.

Kelley, Kevin W. *The Home Planet.* New York: Addison-Wesley, 1988.

Kessenides, Dimitra. "Questions for Stephen Gorevan on Building a Better Robot." *New York Times,* June 17, 2001. <www.nytimes.com/2001/06/17/magazine/17QUESTIONS.html>.

King, Ledyard. "$68B Wastes Away in Rush-Hour Delays, Report Says." *USA Today,* June 20, 2002.

Kinget, G. Marian. *The Master-Myth of Modern Society: A Sketch of the Scientific Worldview and Its Psychosocial Effects.* Lanham, Md.: University Press of America, 2000.

Kircher, Athanasius. *Ars magna lucis et umbrae* (The great technique of light and shadow; 1646). Illus. 78 in *Athaneus Kircher: A Renaissance Man and the Quest for Lost Knowledge,* by Jocelyn Godwin. 83. New York: Thames and Hudson, 1979.

Kleist, Heinrich von. "On the Marionette Theatre" (1811). In *An Abyss Deep Enough: The Letters of H. Kleist.* Ed. Phillip Miller. New York: E. P. Dutton, 1982.

Kofman, Sarah. *Camera Obscura of Ideology* (1973). Trans. Will Straw. Ithaca, N.Y.: Cornell University Press, 1998.

Kraemer, Ross, William Cassidy, and Susan L. Schwartz. *Religions of Star Trek.* Cambridge, Mass.: Westview Press, 2001.

Kroker, Arthur. *The Will to Technology and the Culture of Nihilism: Heidegger, Nietzsche, and Marx.* Toronto: University of Toronto Press, 2004.

Krugman, Paul. "Nation in a Jam." *New York Times,* May 13, 2001, 13.

Kübler-Ross, Elisabeth. *On Death and Dying.* New York: Macmillan, 1969.

Kuhn, Thomas S. *The Structure of Scientific Revolutions.* Chicago: University of Chicago Press, 1970.

Kumar, Krishan. "Utopian and Anti-Utopia in the Twentieth Century." In *Utopia: The Search for the Ideal Society in the Western World.* Ed. Roland Schaer and Lyman Sargent. 2–18. New York: New York Public Library, 2000.

"Kursk Submarine Raised." CNN News, October 8, 2001. <www.cnn.com/2001/WORLD/europe/10>.

Kurzweil, Ray. *The Age of Spiritual Machines*. New York: Penguin/Viking, 1999.

———. "The Evolution of Mind in the Twenty-first Century." In *Are We Spiritual Machines?* Ed. Jay Richards. 12–55. Seattle: Discovery Institute, 2002.

Lame Deer, John (Fire), and Richard Erdoes. *Lame Deer Seeker of Visions*. New York: Simon and Schuster, 1975.

La Mettrie, Julien Offray de. *Man a Machine* (1748). Trans. Gertrude Bussy, M. W. Calkins, and M. Carret. LaSalle, Ill.: Open Court, 1953.

Latour, Bruno. *We Have Never Been Modern*. Trans. Catherine Porter. Cambridge, Mass.: Harvard University Press, 1993.

Laufer, Berthold. *The Prehistory of Aviation*. Chicago: Field Museum of Natural History Press, 1928.

Leary, Warren. "A New, and Ambitious, Home Away from Home." *New York Times*, November 3, 2000, A1.

Leiss, William. *The Domination of Nature*. New York: Brasillier, 1972.

———. *Under Technology's Thumb*. Montreal: McGill–Queen's University Press, 1990.

Levin, David Michael. *The Opening of Vision: Nihilism and the Postmodern Situation*. New York: Routledge, 1988.

Levy-Bruhl, Lucien. *The Notebooks on Primitive Mentality*. Trans. P. Riviere. Oxford: Blackwell, 1979.

———. *Primitive Mentality*. Trans. Lillian Clare. Boston: Beacon, 1923.

Lifton, Robert, and Eric Olson. *Living and Dying*. New York: Praeger, 1974.

Lindbergh, Charles. *The Spirit of St. Louis*. New York: Scribner's and Sons, 1953.

Locke, John. *Essay Concerning Human Understanding* (1690). New York: Dutton, 1976.

London, Herbert, and Albert Weeks. *Myths That Rule America*. Washington, D.C.: University Press of America, 1981.

Lopez, Barry. *Arctic Dreams*. New York: Scribner, 1986.

Lord, Walter. *The Night Lives On*. New York: Morrow/Avon, 1986.

———. *A Night to Remember*. New York: Holt, Rinehart and Winston, 1955.

Love, Sam. "Whatever Became of the Predicted Effortless World?" *Smithsonian*, November 1979, 86–91.

Luke, Timothy W. "Reproducing Planet Earth? The Hubris of Biosphere 2." *The Ecologist* 25 (July/August 1995): 157–62.

Lyall, Sarah. "Man Who Would Be God: Giving Robots Life." *New York Times*, February 2, 2002. <www.nytimes.com/2002/02/02/arts/02GRAN.html>.

MacDonald, Anne L. *Feminine Ingenuity: How Women Inventors Changed America*. New York: Ballantine, 1992.

MacKenzie, Donald, and Judy Wajcmanm, eds. *The Social Shaping of Technology*. Philadelphia: Open University Press, 1999.

Macy, Joanna. *Despair and Personal Power in the Nuclear Age*. Philadelphia: New Society Publishing, 1983.

Marinetti, F. T. "The Founding Manifesto of Futurism" (1909). In *Futurist Manifestos*. Ed. Ubro Apollonio. Trans. R. W. Flint. 19–24. New York: Viking Press, 1970.

———. "The New Religion-Morality of Speed" (1916). In *Marinetti: Selected Writings*. Ed. R. W. Flint. Trans. R. W. Flint and Arthur Coppotelli. 94–96. (New York: Farrar, Straus, and Giroux, 1971.

———. "Portrait of Mussolini" (1929). In *Marinetti: Selected Writings*. Ed. R. W. Flint. Trans. R. W. Flint and Arthur Coppotelli. 158–59. New York: Farrar, Straus, and Giroux, 1971.

Markoff, John. "Happy Birthday HAL: What Went Wrong?" *New York Times,* January 12, 1992, WK 5.

Marsh, Ed W. *James Cameron's Titanic.* New York: HarperCollins, 1997.

The Matrix. Dir. Wachowksi Brothers. Warner Brothers, 1999.

Maxson, Greg. "Second Chance for Biosphere." *Popular Science,* April 1997, 56–60.

McCurdy, Howard. *Space and the American Imagination.* Washington, D.C.: Smithsonian Institution Press, 1997.

McGuigan, Cathleen. "Requiem for an American Icon." *Newsweek,* September 24, 2001, 87.

Medawar, Peter. *The Hope of Progress.* London: Methuen, 1972.

Meikle, Jeffrey. *Twentieth Century Limited: Industrial Design in America, 1925–1939.* Philadelphia: Temple University Press, 1979.

Melville, Herman. *Moby Dick* (1851). Ed. Harrison Hayford and Hershel Parker. New York: W. W. Norton, 1967.

Menzel, Peter, and Faith D'Aluisio. *Robo Sapiens: Evolution of a New Species.* Cambridge: Massachusetts Institute of Technology Press, 2000.

Merchant, Caroline. *The Death of Nature: Women, Ecology, and the Scientific Revolution.* New York: Harper and Row, 1980.

Meredith, Robyn. "In Detroit, a Sex Change." *New York Times,* May 16, 1999, WK 3.

Metropolis. Dir. Fritz Lang. Janus Films, 1923.

Michel, Jean. *Dora.* New York: Holt, Rinehart, and Winston, 1975.

Midgley, Mary. *Science as Salvation: A Modern Myth and Its Meaning.* New York: Routledge, 1992.

Miller, Christopher. "Gadget Fetish." *Popular Science,* January 2001, 58–61.

Miller, David L. *The New Polythesim: Rebirth of the Gods and Goddesses.* New York: Harper and Row, 1974.

Mingo, Jack. *How the Cadillac Got Its Fins.* New York: Harper, 1994.

Minsky, Marvin. "R.U.R. Revisited." In *Visions of Technology.* Ed. Richard Rhodes. 369. New York: Simon and Schuster, 1999.

———. *The Society of Mind.* New York: Simon and Schuster, 1985.

———. "Why People Think Computers Can't." In *Computers, Ethics, and Society.* Ed. M. David Ermann, Mary B. Williams, and Claudio Gutierrez. 145–66. New York: Oxford, 1990.

Mitchell, Edgar. *The Way of the Explorer.* New York: Putnam's, 1996.

Mohawk, John C. *Utopian Legacies.* Santa Fe, N.M.: Clear Light, 2000.

Montville, Leigh. *At the Altar of Speed.* New York: Doubleday, 2001.

Moore, Frazier. "Watson Unravels DNA Mystery." *Ithaca Journal,* January 1, 2004, 7A.

Moore, Thomas. "James Hillman: Psychology with Soul." *Religious Studies Review* 6.4 (October 1980): 278–86.

———. *The Re-enchantment of Everyday Life.* New York: HarperCollins, 1996.

Moravec, Hans. *Robot: Mere Machine to Transcendent Mind.* Oxford: Oxford University Press, 1999.

More, Thomas. *Utopia* (1516). Ed. and trans. Robert Adams. New York: Norton, 1975.

Morris, R. G. "The Magic Lantern in 1693." *New Magic Lantern Journal* 1.1 (April 1978): 3.

———. "Victorian Youth and the Magic Lantern." *New Magic Lantern Journal* 2.1 (January 1981): 2–3.

"Move Over, Asimo." *Ithaca Journal,* January 8, 2003, 1E.

Mumford, Lewis. *The Myth of the Machine: Technics and Human Development.* New York: Harcourt, Brace, and World, 1966.

————. *The Myth of the Machine: The Pentagon of Power.* New York: Harcourt, Brace, and World, 1970.

————. *Technics and Civilization* (1934). New York: Harcourt, Brace, and World, 1963.

Naisbitt, John, with Naha Nesbitt and Douglas Philips. *High Tech High Touch.* New York: Broadway Books, 1999.

Naudet, A. "La lanterne magique" (The magic lantern; 1826). *New Magic Lantern Journal* 2.2 (January 1982): 8–9.

Naughton, Keith. "Green and Mean." *Newsweek,* November 22, 2004, 50–56.

Nelkin, Dorothy. *Selling Science: How the Press Covers Science and Technology.* New York: W. H. Freeman, 1987.

Nelson, Victoria. *The Secret Life of Puppets.* Cambridge: Harvard University Press, 2001.

Nietzsche, Friedrich. "From *The Gay Science.*" In *The Portable Nietzsche.* Ed. Walter Kaufmann. 93–102. New York: Viking Press, 1968.

————. "Truth and Illusion." In *The Philosophy of Nietzsche.* Ed. Geoffrey Clive. 503–15. New York: New American Library, 1965.

————. *The Will to Power.* Ed. Walter Kaufmann. New York: Vintage, 1967.

"Noah's Ark: The Sequel." *New York Times Magazine,* January 31, 1988, 62.

Noble, David F. *The Religion of Technology.* New York: Knopf, 1997.

Nordland, Rod. "Where Is the Next Chernobyl?" *Newsweek,* October 18, 1999, 32–40.

Norman, Donald A. *Emotional Design: Why We Love (or Hate) Everyday Things.* New York: Basic Books, 2004.

"Notable Marine Disasters." *Boston Globe,* April 16, 1912.

Nye, David. *American Technological Sublime.* Cambridge: Massachusetts Institute of Technology Press, 1994.

O'Connell, Jeffery, and Arthur Myers. *Safety Last: An Indictment of the Auto Industry.* New York: Random House, 1966.

"Officers Astounded at Loss of the *Titanic.*" *Boston Globe,* April 16, 1912, 6.

O'Neill, Gerard. *The High Frontier: Human Colonies in Space.* New York: Morrow, 1977.

Orwell, George. *1984.* New York: Harcourt, Brace, Janovitch, 1949.

Ovid. *Metamorphoses.* Trans. Rolfe Humphries. Bloomington: Indiana University Press, 1955.

Pacey, Arnold. *The Culture of Technology* (1983). Cambridge: Massachusetts Institute of Technology Press, 1994.

————. *The Maze of Ingenuity: Ideas and Idealism in the Development of Technology.* Cambridge: Massachusetts Institute of Technology Press, 1974.

————. *Meaning in Technology.* Cambridge: Massachusetts Institute of Technology Press, 1999.

————. *Technology in World Civilization.* Cambridge: Massachusetts Institute of Technology Press, 1992.

Panofsky, Erwin. "Die perspektive als 'symbolishe form'" (Perspective as "symbolic form"). *Vortrage der Bibliothek Warburg* (1924–25): 258–354. See also *Perspective as Symbolic Form.* Trans. Christopher Wood. New York: Zone Books, 1991.

Papanek, Victor. *Design for the Real World.* New York: Pantheon/Bantam, 1972.

Park, Robert L. *Voodoo Science.* New York: Oxford University Press, 2000.

Passell, Peter. "Our Man in a MIG." *New York Times Magazine,* December 19, 1993, 64–65.

Patton, Phil. *Made in USA.* New York: Grove Weidenfeld, 1992.

————. "A Proud and Primal Roar." *New York Times,* January 12, 2003, 1–4.

Perera, Sylvia Brinton. *Celtic Queen Maeve and Addiction: An Archetypal Perspective.* York Beach, Me.: Nicholas-Hays, 2001.

————. *Descent to the Goddess: A Way of Initiation for Women.* Toronto: Inner City Books, 1981.

Perlmutter, Dawn. "The Sacrificial Aesthetic: Blood Rituals from Art to Murder." *Anthropoetics* 5.2 (Fall 1999/Winter 2000). <www/anthropoetics.ucla.edu>.

Perrow, Charles. *Normal Accidents: Living with High-Risk Technologies.* New York: Basic Books, 1984.

Pfarr, Anastasia, ed. "Ashcroft vs. Greenpeace." *Greenpeace Update* (Winter 2003): 2–6.

Pinocchio. Dir. Walt Disney. Walt Disney Home Video. 1940.

Piszkiewicz, Dennis. *The Nazi Rocketeers: Dreams of Space and Crimes of War.* London: Praeger, 1995.

Plato. "Gorgias." In *Collected Dialogues.* Ed. Edith Hamilton and Huntington Cairns. 229–307. Princeton, N.J.: Princeton University Press, 1978.

————. "Phaedo." In *Collected Dialogues.* Ed. Edith Hamilton and Huntington Cairns. 40–98. Princeton, N.J.: Princeton University Press, 1978.

Pollack, Robert. *The Missing Moment: How the Unconscious Shapes Modern Science.* Boston: Houghton Mifflin, 1999.

Porta, John Baptista (Giambattista Della Porta). *Natural Magick* (1658). New York: Basic Books, 1957.

Postman, Neil. *Technopoly: The Surrender of Culture to Technology.* New York: Random House, 1992.

Preston, Julia. "A Fatal Case of Fatalism: Mexico City's Air." *New York Times,* February 14, 1999, NWR 3.

Rapaille, Clothaire. "Archetype Discoveries Worldwide." January 2003. <www.rapaille institute.com>.

Ray, Thomas. "Kurzweil's Turing Fantasy." In *Are We Spiritual Machines?* Ed. Jay Richards. 116–27. Seattle: Discovery Institute, 2002.

Recer, Paul. "Oxygen Depletion Doomed Biosphere 2." *Ithaca Journal,* November 15, 1996, 4B.

Reed, Hayes. "SUV Explosion." *Ithaca Times,* July 6, 2000, 9–11.

Return of the Jedi. Dir. Richard Marquand. Lucasfilms, 1983.

Rhodes, Richard, ed. *Visions of Technology.* New York: Simon and Schuster, 1999.

Robertson, Etienne. *Memoires: Recreatifs scientifiques et anecdotiques* (Memories: Amusing, scientific, and recreational). Paris, 1831.

Robocop. Dir. Paul Verhoeven. Orion Pictures, 1987.

Romanyshyn, Robert. *Technology as Symptom and Dream.* London: Routledge, 1989.

Rosenberg, Nathan, et al., eds. *Technology and the Wealth of Nations.* Stanford, Calif.: Stanford University Press, 1992.

Roszak, Theodore. *Where the Wasteland Ends.* Garden City, N.Y.: Doubleday Anchor, 1972.

Rothenberg, David. *Hand's End: Technology and the Limits of Nature.* Berkeley: University of California Press, 1993.

Rothschild, Joan, ed. *Machina ex Dea: Feminist Perspectives on Technology.* New York: Pergamon Press, 1983.

Royce, Joseph R. *The Encapsulated Man.* Princeton, N.J.: Van Nostrand, 1964.

Russell, Colin A. *The History of Valency.* Leicester: Leicester University Press, 1971.

Russell, Bertrand. *A History of Western Philosophy.* New York: Simon and Schuster, 1945.

Sardello, Robert. *Facing the World with Soul.* Hudson, N.Y.: Lindisfarne, 1992.

Sartre, Jean-Paul. *Being and Nothingness* (1943). Trans. Hazel Barnes. New York: Gramercy Books, 1956.

————. *No Exit* (1945). Trans. S. Gilbert. New York: Random House, 1955.

Sawyer, Kathy. "NASA Oversight Panelists Resign: Departures Roil Safety Reform Efforts." *Washington Post,* September 24, 2003, A3.

Schnarrs, Steven. *Megamistakes: Forecasting and the Myth of Rapid Technological Change.* New York: Macmillan, 1989.

Schumer, F., and J. Zubin. "Projective Techniques." In *Encyclopedia of Psychology.* 833. New York: Seabury, 1979.

Schwartz, Heinrich. "Art and Photography: Forerunners and Influences." *Magazine of Art* 42 (November 1949): 252–57.

Schwartz, John. "A Robot That Works in the City Sewer." *New York Times,* March 8, 2001. <www.nytimes.com/2001/03/08/technology/08ROBO.html>.

Scott, A. O. "A.I.: Do Androids Long for Mom?" *New York Times,* June 29, 2001. <www.nytimes.com/2001/06/29ARTI.html>.

Searle, John. "I Married a Computer." In *Are We Spiritual Machines?* Ed. Jay Richards. 56–77. Seattle: Discovery Institute, 2002.

Segal, Harold. *Pinocchio's Progeny.* Baltimore: Johns Hopkins University Press, 1995.

Segal, Howard. *Technological Utopianism in American Culture.* Chicago: University of Chicago Press, 1985.

Seidel, Peter. *Invisible Walls: Why We Ignore the Damage We Inflict on the Planet . . . and Ourselves.* Amherst, N.Y.: Prometheus Books, 2001.

Shakespeare, William. *Henry VI Part 2.* In *The Collected Works of William Shakespeare.* 600–36. New York: P. F. Collier, 1925.

————. *A Midsummer Night's Dream.* In *The Collected Works of William Shakespeare.* 167–90. New York: P. F. Collier, 1925.

Short Circuit. Dir. John Badham. Tri-Star Pictures, 1986.

Simon, Herbert. "Culture to the Nth Power." In *Visions of Technology.* Ed. Richard Rhodes. 246–52. New York: Simon and Schuster, 1999.

Skolimowski, Henryk. *The Participatory Mind.* London: Penguin/Arkana, 1994.

Smith, Paul. *The Robot Exhibit.* New York: American Craft Museum, 1984.

Smollett, Tobias. *Ferdinand, Count Fathom* (1753). New York: Greenberg, 1926.

Snell, Joel. "Predictions: Robotic Sex." In *Visions of Technology.* Ed. Richard Rhodes. 371–72. New York: Simon and Schuster, 1999.

"Sony's Humanoid Robot Learns to Jog." CNN News, December 12, 2004. <http://www.cnn.com/2003/TECH/pttech/12/18/SONY.robot.ap>.

Sobel, Dava. *Galileo's Daughter: A Historical Memoir of Science, Faith, and Love.* New York: Walker and Co., 1999.

SOS Titanic. Dir. Billy Hade. EMI Films, 1979.

Space Cowboys. Dir. Clint Eastwood. Warner Brothers Pictures, 2000.

Spignesi, Stephen J. *The Complete Titanic.* Secaucus, N.J.: Birch Lane/Carol Publishing, 1998.

Spotila, James. "The Tragic Oil Spill Was Preventable." *Philadelphia Enquirer,* December 7, 2004. <http://www.Philly.com/mld/inquirer/news/opinion/local2>.

Stahl, William. *God and the Chip: Religion and the Culture of Technology.* Waterloo, Ont.: Wilfred Lauier Press, 1999.

Standage, Tom. *The Turk.* New York: Penguin, 2002.

Star Trek: First Contact. Dir. Jonathan Frakes. Paramount, 1996.

Star Wars. Dir. George Lucas. Twentieth-Century Fox, 1977.

Stepford Wives. Dir. Brian Forbes. Paramount, 1975.

Steward, Ben. "Spirit of the Salt." *Popular Mechanics,* January 2001, 72–77.

Stewart, J. A. *The Myths of Plato.* New York: Macmillan, 1905.

Stivers, Richard. *Technology as Magic*. New York: Continuum, 1999.

Stone, Brad. "Attack of the Bots." *Newsweek,* May 28, 2001, 40–41.

———. *Gearheads: The Turbulent Rise of Robotic Sports*. New York: Simon and Schuster, 2003.

Stone, Jennifer. "Pinocchio and Pinocchiology." *American Imago: Studies in Psychoanalysis and Culture* 51.3 (Fall 1994): 329–42.

Stover, Dawn. "Inside Biosphere II." *Popular Science,* November 1990, 54–112.

Stulinger, Ernst, and Frederick Ordway. *Wernher von Braun, Crusader for Space: A Biographical Memoir*. Malabar, Fla.: Krieger Publishing, 1994.

Suplee, Curt. "Robot Revolution." *National Geographic,* July 1997, 76–95.

"Surveillance Cameras Incite Protest." *New York Times,* July 16, 2001. <www.nytimes.com/2001/07/16/16Tamp.html>.

Temkin, Owsei. "Health and Disease." In *The Dictionary of the History of Ideas*. Vol. 2. Ed. Philip P. Weiner. 395–407. New York: Scribner's, 1973.

Tenner, Edward. *Why Things Bite Back: Technology and the Revenge of Unintended Consequences*. New York: Knopf, 1997.

The Terminator. Dir. James Cameron. Orion Pictures, 1984.

The Terminator 2. Dir. James Cameron. Carolco Pictures, 1991.

Texas Transportation Institute. *2004 Urban Mobility Study*. Texas A&M University. January 2005. <www.mobility.edu/2001/news_release>.

Tillich, Paul. *Systematic Theology*. Chicago: University of Chicago Press, 1967.

Titanic. Dir. James Cameron. Paramount, 1997.

Total Recall. Dir. Paul Verhoeven. Carolco Pictures, 1990.

Totilo, Stephen. "Letting Gamers Play God, and Now Themselves." *New Tork Times*. Sept. 9, 2002, A2

Trimble, Bjo. *Star Trek Concordance*. New York: Citadel Press, 1995.

Tron. Dir. Steven Lisberger. Disney, 1982.

Tuan, Yi-Fu. *Segmented Worlds and Self*. Minneapolis: University of Minnesota Press, 1982.

Turing, Alan. "Computing Machinery and Intelligence." *Mind: A Quarterly Review of Psychology and Philosophy* 59.236 (October 1950): 433–60.

Turkle, Sherry. *Life on the Screen: Identity in the Age of the Internet*. New York: Touchstone/Simon and Schuster, 1995.

———. *The Second Self: Computers and the Human Spirit*. New York: Simon and Schuster, 1984.

2001: A Space Odyssey. Dir. Stanley Kubrick. MGM Pictures, 1968.

Ullman, Ellen. *Close to the Machine: Technophilia and Its Discontents*. San Francisco: City Lights Books, 1997.

———. "Programming the Post-Human." *Harper's,* October 2002, 60–70.

Vallemont, M. de. *Le physique occult* (Occult physics). Amsterdam: N.p., 1693.

Vergano, Dan. "Brave New World of Biosphere 2." *Science News,* November 16, 1996, 312–13.

Vistica, Gregory. "Seeing through Stealth." *Newsweek,* July 5, 1999, 30.

Vizard, Frank. "Raising the Kursk." *Popular Science,* April 2001, 42–47.

Volkmann, Wilhelm. *Lehrbuch der psychologie* (Psychology textbook). 2 vols. Cothen: Schultze, 1876.

Von Franz, Marie-Louise. *Projection and Recollection in Jungian Psychology*. Trans. W. Kennedy. La Salle, Ill.: Open Court, 1980.

Wallace, William A. "Experimental Science and Mechanics." In *Dictionary of the History of Ideas*. Vol. 2. Ed. Philip P. Weiner. 202. New York: Scribner's, 1973, 196–207.

Wallace, B. Alan. *The Taboo of Subjectivity.* Oxford: Oxford University Press, 2000.

Wallace, Bruce, and Angela Doland. "Concorde." *Montreal Gazette,* July 26, 2000, A1–5.

Walter, William J. *Space Age.* New York: Random House, 1992.

Waltz, David. "Artificial Intelligence." *Scientific American,* October 1982, 118–33.

Warrick, Patricia. *The Cybernetic Imagination in Science Fiction.* Cambridge: Massachusetts Institute of Technology Press, 1980.

Warshall, Peter. "Lessons from Biosphere 2." *Whole Earth Review* (Spring 1996): 22–27.

Weber, Max. "Science as a Vocation." In *From Max Weber: Essays in Sociology.* Ed. and trans. H. H. Gerth and C. Wright Mills. 129–56. New York: Oxford University Press, 1946.

Weiner, Tim. "How to Build Weapons When Money Is No Object." *New York Times,* April 16, 2000, WK 3.

Wells, H. G. *First Men on the Moon* (1901). New York: Oxford University Press, 1995.

———. *Men Like Gods.* New York: Cassell, 1923.

———. *A Modern Utopia* (1905). Lincoln: University of Nebraska Press, 1967.

———. *War of the Worlds* (1898). New York: Airmont Publishing, 1964.

———. *The Wheels of Chance and Time Machine* (1895). New York: Dutton, 1965.

———. *When the Sleeper Awakes.* New York: Harper, 1899.

Westworld. Dir. Michael Crichton. MGM Pictures, 1973.

Wheeler, Larry. "Boehlert Applauds Bush Moon Plans." *Ithaca Journal,* January 10, 2004, B3.

White, Lynn. "The Historical Roots of Our Ecological Crisis." *Science* 15 (March 10, 1967): 1203–7.

Whitman, Walt. "Song of Myself" (1855). In *American Poetry and Prose.* Vol. 2. Ed. Norman Foerster. 852–85. Boston: Houghton Mifflin, 1957.

Whynott, Douglas. "Machines That Walk, Talk, Move, and Show Humanlike Emotions Are No Longer Science Fiction." *Discover,* October 1999, 66–73.

Whyte, L. L. *The Unconscious before Freud.* New York: Basic Books, 1960.

Wier, Dennis. *Trance: From Magic to Technology.* Ann Arbor, Mich.: Trans-Media, 1996.

Wilson, Jim. "Recycling Biosphere 2." *Popular Mechanics,* August 1997, 38–39.

Winner, Langdon. *Autonomous Technology: Technics-Out-of-Control as a Theme in Political Thought.* Cambridge: Massachusetts Institute of Technology Press, 1977.

———. *The Whale and the Reactor: A Search for Limits in an Age of High Technology.* Chicago: University of Chicago Press, 1986.

The Wizard of Oz. Dir. Victor Fleming. MGM Pictures, 1939.

Wolfe, Gary K. *The Known and the Unknown: The Iconography of Science Fiction.* Kent, Ohio: Kent State University Press, 1979.

Wolfgang, Lon. "Biosphere 2 Turned Over to Columbia University" *Science,* November 17, 1995, 1111.

Wolinsky, Stephen. *Trances People Live.* Las Vegas: Bramble, 1991.

Wunderlich, Richard, and Thomas Morrissey. *Pinocchio Goes Postmodern: Perils of a Puppet in the United States.* New York: Routledge, 2002.

Yenne, Bill. *The World's Worst Aircraft.* Greenwich, Conn.: Dorset Press, 1990.

Zera, Thomas. *Ore-Oil-Bulk: Pictoral History of Bulk Shipping Losses during the 1980s.* Bethel, Conn.: Routledge, 1996.

Zimmerman, Michael. *Heidegger's Confrontation with Modernity: Technology, Politics, Art.* Bloomington: Indiana University Press, 1990.

INDEX

LEE WORTH BAILEY is associate professor of religion and culture at Ithaca College in Ithaca, New York. He graduated in industrial design at the University of Illinois (BFA), in Christianity at Union Theological Seminary at Columbia University (M.Div.), and in the Humanities Interdisciplinary Program at Syracuse University (Ph.D). He is listed in *Who's Who in the World* and other *Who's Who* directories and has published many articles on topics such as projection in the psychology of religion, Dutch approaches to projection theory, the history of the camera obscura, the magic lantern, and near-death experiences. He is the editor with Jenny Yates of *The Near-Death Experience: A Reader* (1996) and with Mary Pat Fisher of *An Anthology of Living Religions* (2000). He was the video editor for a DVD-ROM, *The Sacred World: Encounters with the World's Religions*. He is series editor for *Introduction to the World's Major Religions* (2005) and the author of the volume on Christianity. He is a member of the American Academy of Religion and former copresident of its Eastern International Region. He is the former chair of the Department of Philosophy and Religion at Ithaca College.

The University of Illinois Press
is a founding member of the
Association of American University Presses.

University of Illinois Press
1325 South Oak Street
Champaign, IL 61820-6903
www.press.uillinois.edu